CIRIA SPECIAL PUBLICATION 78
PSA CIVIL ENGINEERING TECHNICAL GUIDE 60

1991

GW00725415

Building on derelict land

B A Leach BSc(Eng) CEng FICE FIStructE FASCE FGS

H K Goodger BSc(Eng) CEng MICE

PSA **Specialist Services**

Apollo House
36 Wellesley Road
Croydon CR9 3RR
Telephone 081-760 8646

CIRIA

CONSTRUCTION INDUSTRY RESEARCH AND INFORMATION ASSOCIATION
6 STOREY'S GATE, LONDON SW1P 3AU   TELEPHONE 071-222 8891

Summary

Detailed guidance is given on the investigation, appraisal, remedial treatment and building on sites which have become derelict, mainly as a result of human activity. The body of the publication is in two parts; Part A outlines the basic principles and procedure of the development of derelict sites and Part B explains in detail the problems posed by derelict sites and how solutions can be approached.

The hazards of filled and contaminated sites are discussed with reference to specific industrial or other past uses, which may be identifiable from historical records. The initial assessment, detailed site investigation, sampling techniques and final appraisal of a derelict site are described. Remedial treatment of filled and contaminated sites is covered and design of foundations, below-ground services, roads and other paving is discussed. The value of monitoring and the importance of contractual aspects are stressed. Detailed guidance is given on matters concerning the safety of site workers and third parties.

Four appendices cover funding of development, application to Scotland, the application of general principles to specific sites and gas-sampling procedures.

LEACH, B.A. and GOODGER, H.K.
Building on derelict land
Construction Industry Research and Information Association
Special Publication 78, 1991

Keywords: (*from Construction Industry Thesaurus*)
Derelict land, industrial dereliction, spoil heaps, reclamation of derelict land, restoration of derelict land

Reader Interest:
Engineers, developers, planners, environmental scientists

© CIRIA, 1991

ISBN: 0 86017 323 2

Printed in Great Britain
at The Alden Press,
Oxford

		CLASSIFICATION
AVAILABILITY		Unrestricted
CONTENT		Guidance on best current practice
STATUS		Committee-guided
USER		Engineers

Foreword

This Special Publication was prepared under contract for PSA Specialist Services (formerly the Directorate of Civil Engineering Services of the Property Services Agency, DoE). It is one of a series which gives guidance on specific aspects of foundation engineering. These publications are for use in the PSA Specialist Services, and have also been disseminated to industry.

Following usual CIRIA procedures, a Steering Group was established, from a broad range of construction interests and related scientific disciplines, to ensure that the guidance given is representative of current industrial practice. The Steering Group comprised:

J. May (*Chairman*)	Property Services Agency
F.R.D. Chartres (*Chairman following Mr May's retirement through ill health*)	PSA Specialist Services
D.L. Barry	W.S. Atkins Planning Consultants
M.J. Beckett	Department of the Environment
J.A. Charles	Building Research Establishment
R.W. Johnson	National House-Building Council
G.W. Lowe	Carpenter and Lowe Ltd
I.H. McFarlane	Mitchell, McFarlane & Partners
D.J. Palmer	Consultant
A. Parker	Harwell Laboratory
M.A. Smith	Clayton, Bostock, Hill & Rigby
G.H. Thomson	Cementation Piling & Foundations Ltd
S. Thorburn/N.W. Buchanan	Thorburn Associates

The text was written by B.A. Leach and H.K. Goodger of Allott & Lomax, Consulting Engineers.

This publication is intended as a guide to current UK procedures and techniques for the investigation, appraisal, reclamation and building development of sites which have been made derelict mainly by human activity.

The basic principles for the treatment of derelict land are becoming established, but development of derelict sites is not an exact science, nor is it ever likely to be so. Engineers responsible for such development, therefore, not only have to exercise considerable judgement but also to seek the support of professionals from other disciplines.

The publication is in two Parts: Part A outlines the whole process of developing derelict sites; Part B provides detailed guidance. The guidance is primarily intended for engineers who already have sound experience in ground engineering for greenfield developments but who are unfamiliar with the special problems posed by dereliction. Others concerned with these sites, as developers, planners, construction professionals or environmental scientists, will find the publication useful; particularly for the overview in Part A, but also for the explanations detailed in Part B.

Acknowledgements

CIRIA and the authors are grateful to the many experts who have generously given advice and help during the research for this work, and in particular:

W. Baker	City of Birmingham Industrial Research Laboratories
P.F. Beever	Fire Research Station
R.M. Bell	Environmental Advisory Unit, Liverpool University
G. Brownsword	Divisional EHO, Dudley Metropolitan Borough
T. Cairney	Dean of the Faculty of Construction, Liverpool Polytechnic
R.J. Carpenter	Carpenter and Lowe Ltd
P. Cooney	London Borough of Greenwich
J. Crowther	Ward Ashcroft Ltd
T.W.A. Durn	Principal EHO, London Borough of Hillingdon
A.C. Ellis	Clayton, Bostock, Hill & Rigby
W.H. Harrison	Building Research Establishment
C. Knipe	Johnson, Poole and Bloomer
N. Lepp	Department of Biology, Liverpool Polytechnic
M.P. Moseley	Keller Foundations Ltd
R. Sharman	Commission for the New Towns, Corby, Northants.
C. Stonehouse	Midland and General Homes Ltd
D.J. Taylor	Department of Development and Planning (Derelict land), Merseyside County Council

This project developed from earlier work on behalf of CIRIA by A.J. Weltman. The work of D. Oldham, then of Allott & Lomax, in the early stages of this project is also gratefully acknowledged.

CIRIA's Research Manager for the project was F.M. Jardine.

Cover photograph   The authors and publishers are grateful to W.S. Atkins Consultants Ltd for permission to reproduce the photograph on the front cover: Advance works on former paintworks site for the reclamation of Caledonian Wharf, London Docklands, for residential development.

Client: London Docklands Development Corporation
Consultant (Engineering and Ground Contamination): W.S. Atkins Consultants Ltd
Contractor: G. Percy Trentham Ltd.

Contents

	Page
SUMMARY	2
FOREWORD	3
ACKNOWLEDGEMENTS	4
LIST OF TABLES	9
LIST OF FIGURES	10
LIST OF ABBREVIATIONS	11
GLOSSARY	12

PART A

1. THE PRESENT SCENE	15
1.1 Fill sites	16
1.2 Industrially contaminated sites	17
1.3 Essentials and implicit responsibilities of development	17
2. LOGICAL PROCEDURE AND STRUCTURE OF THIS PUBLICATION	19
3. TECHNICAL PROBLEMS AND PROFESSIONAL EXPERTISE	21
3.1 Sites and hazards	21
3.2 Hazards and end uses	21
3.3 Professional expertise	22
4. POTENTIAL CONSTRAINTS	23
4.1 Technical and environmental	23
4.2 Statutory	23
4.3 Safety	24
4.4 Time	24
4.5 Cost	24
4.6 Consequence of constraints	25
5. INVESTIGATION STRATEGY	26
5.1 Scope	26
5.2 Phase 1: initial assessment of site	26
5.3 Phase 2: detailed site investigation	27
5.4 Safety in site investigation	28
6. SITE APPRAISAL	29
6.1 Scope	29
6.2 Acceptance criteria for hazard levels	29
6.3 Options for action	29
7. REMEDIAL TREATMENT OF GROUND	31
7.1 Fill sites	31
7.2 Contaminated sites	32
7.3 Special considerations	32
7.4 Necessary and routine works common to all derelict sites	32
8. DESIGN MEASURES	34
9. MONITORING	35
10. CONTRACTUAL CONSIDERATIONS	36

PART B

11. ENGINEERING PROBLEMS OF FILL | 37
    11.1   Type and content of fill sites | 37
    11.2   Problems caused by fill site hazards | 38
    11.3   Engineering behaviour of fill | 38
    11.4   Problems of slags, mining spoils and building wastes | 44

12. HAZARDS OF CONTAMINATION | 48
    12.1   Type and content of contaminated sites | 48
    12.2   Problems from contaminated site hazards | 48
    12.3   Obstructions | 49
    12.4   Health and environmental hazards: general | 50
    12.5   Health hazards: inhalation | 52
    12.6   Health hazards: ingestion | 54
    12.7   Health hazards: contact with contaminants | 56
    12.8   Phytotoxicity | 57
    12.9   Combustion | 58
    12.10 Fire and explosion | 60
    12.11 Gas emission | 61
    12.12 Aggressive attack and corrosion | 63
    12.13 Leachate and contamination of groundwater | 63
    12.14 Migration of contaminants | 64
    12.15 Biodegradation of refuse fills | 66

13. CONSTRAINTS ON DEVELOPMENT | 67
    13.1   Technical and environmental | 67
    13.2   Statutory | 69
    13.3   Safety | 71
    13.4   Time | 71
    13.5   Cost | 71

14. INITIAL ASSESSMENT | 73
    14.1   Multi-disciplinary approach | 73
    14.2   Desk study | 74
    14.3   Site reconnaissance | 77
    14.4   Monitoring | 79
    14.5   Information for planning detailed site investigation | 79
    14.6   Report on initial assessment | 80

15. DETAILED SITE INVESTIGATION | 81
    15.1   Objective | 81
    15.2   Standards and workmanship | 81
    15.3   Monitoring | 81
    15.4   Preliminary works | 81
    15.5   Exploration methods of site investigation | 82
    15.6   Sampling strategy for contamination | 87
    15.7   Chemical analyses | 90
    15.8   Groundwater investigation | 93
    15.9   Combustibility investigation | 94
    15.10 Investigation of slags, mining spoils and building wastes | 95
    15.11 Investigation of landfill gas | 96
    15.12 Reporting | 96

16. SAMPLING TECHNIQUES | 97
    16.1   Sampling of solids | 97
    16.2   Sampling of liquids | 98
    16.3   Sampling of gases | 100
    16.4   Sampling biological matter | 101
    16.5   Sampling combustible materials | 101
    16.6   Sampling slags and other expansive fills | 101
    16.7   Sampling radioactivity | 101

CIRIA Special Publication 78

17. DETAILED SITE APPRAISAL 102
    17.1   Acceptance criteria 102
    17.2   Options and decision making 102
    17.3   Appraisal of fill 104
    17.4   Appraisal of health hazard 116
    17.5   Appraisal of combustibility 119
    17.6   Appraisal of aggressive attack and corrosion hazard 121
    17.7   Appraisal of slags, mining spoils and building wastes used as fill 123
    17.8   Appraisal of gas-emission hazard 124
    17.9   Appraisal of leachate 128

18. REMEDIAL TREATMENT OF FILL SITES 130
    18.1   Choice of technique 130
    18.2   Pre-loading 131
    18.3   Excavation and compaction or replacement 132
    18.4   Vibroflotation and vibrated stone columns 133
    18.5   Dynamic compaction 136
    18.6   Remedial treatment of slags and other expansive materials in the ground 138

19. REMEDIAL TREATMENT OF CONTAMINATED SITES 139
    19.1   Use of covers 139
    19.2   Excavation and disposal 145
    19.3   *In-situ* treatment 147
    19.4   Soil processing: on- or off-site 148
    19.5   Control of landfill gas movement 149
    19.6   Treatment of potential fire hazards 151
    19.7   Combating aggressive ground conditions 153
    19.8   Protecting water resources, foundations, etc. against leachate 154
    19.9   Imported fill materials 156

20. DESIGN MEASURES 159
    20.1   Design against settlement 159
    20.2   Design against combustion hazard 165
    20.3   Design of buildings against gas 165
    20.4   Design against aggressive attack and corrosion 169
    20.5   Protection of services 170
    20.6   Design measures against slags and other expansive materials 172

21. MONITORING PROCEDURES 173
    21.1   Need for monitoring 173
    21.2   Contaminant migration 173
    21.3   Groundwater levels and quality 174
    21.4   Leachate movement and quality 174
    21.5   Gas emission 174
    21.6   Ground temperatures 175
    21.7   Settlement monitoring 175

22. CONTRACTUAL GUIDANCE 177
    22.1   Introduction 177
    22.2   Contracts for work to ground level 177
    22.3   Specification 179
    22.4   The tender 180
    22.5   Supervision of construction 181

23. SAFETY IN SITE WORKING 182
    23.1   Introduction 182
    23.2   Stages of development and restricted areas 182
    23.3   Hazards and precautions 183
    23.4   Protective measures 187
    23.5   Safe working procedures 190
    23.6   Containment of contamination 192
    23.7   Further information 193

REFERENCES 195

APPENDICES

1. FINANCIAL ASSISTANCE FOR DEVELOPMENT   205
   A1.1 DoE Derelict Land Grant   205
   A1.2 Eligibility for DLG   205
   A1.3 Current procedure on land improvement   205
   A1.4 Private sector financing   206
   A1.5 Further information   206

2. APPLICATION TO SCOTLAND   207

3. BASIC PRINCIPLES APPLIED TO SPECIFIC TYPES OF SITE   208
   A3.1 Industrial sites   208
   A3.2 Commercial sites   213
   A3.3 Domestic sites   216
   A3.4 Opencast fill sites   217

4. SYSTEMATIC GAS INVESTIGATION   221
   A4.1 Scope of investigation   221
   A4.2 Preliminary investigation   222
   A4.3 Deeper investigation   222
   A4.4 Gas-sampling methods   223
   A4.5 Interpretation of results   224
   A4.6 Long-term gas monitoring   224
   A4.7 Estimation of future gassing potential   224

INDEX   225

List of Tables

Page

Table 1 Potential hazards of derelict sites — 21
Table 2 Sites, hazards and end-use targets — See inside back cover
Table 3 Application of design measures — 34
Table 4 Examples of contaminated sites — 48
Table 5 Some significant contaminants — 49
Table 6 Sensitivity of targets on contaminated sites — 53
Table 7 Characteristics and effects of hazardous gases — 54
Table 8 Contaminants which may affect groundwater — 56
Table 9 Some self-heating materials — 59
Table 10 Properties of some common flammable liquids and gases — 60
Table 11 Composition of typical landfill leachates — 64
Table 12 Summary of sources of information for the appraisal of derelict sites — 76
Table 13 Geophysical methods in surveying fill sites — 86
Table 14 Summary of *in-situ* geotechnical tests for fill — 87
Table 15 Contaminant hazards related to end use — 92
Table 16 Typical potential self-weight settlement of fill materials — 106
Table 17 Compressibility of fills — 107
Table 18 Creep settlement rate parameter, α — 109
Table 19 Tentative 'trigger concentrations' for selected inorganic contaminants — 117
Table 20 Tentative 'trigger concentrations' for contaminants associated with former coal carbonisation sites — 118
Table 21 Potential aggressivity from oxidation of pyrite — 124
Table 22 Techniques for increasing the density of fill — 130
Table 23 Suitability of various fills for vibratory treatment — 134
Table 24 Possible contaminants and relevant actions — 145
Table 25 Suitability of materials as imported fill — 156
Table 26 Preliminary guide to choice of foundation type on fill — 159
Table 27 Materials used for service pipes — 170
Table A1 Industrial works and associated hazards — 208
Table A2 Possible sources of some commonly occurring contaminants — 209
Table A3 Additional contaminant hazards encountered on gasworks sites — 212
Table A4 Commercial works and associated hazards — 214
Table A5 Maximum metal concentrations for three scrapyards — 214
Table A6 Additional contaminant hazards encountered on scrapyard sites — 216

List of Figures

		Page
Figure 1	*Derelict land in England, 1982*	15
Figure 2	*Pear Tree Lane Colliery, Dudley, c.1890*	17
Figure 3	*Layers of old contaminants*	24
Figure 4	*Underground combustion*	28
Figure 5	*Vibro-replacement (stone columns) in fill*	31
Figure 6	*Constituents of domestic waste, 1940 and 1982*	39
Figure 7	*Observed settlement of fills*	40
Figure 8	*Creep settlement of fill under self-weight*	41
Figure 9	*Collapse settlement of opencast backfill due to inundation*	42
Figure 10	*Settlement suffered by building before and after collapse of fill*	43
Figure 11	*Acidity and sulphate production from oxidation of pyrite*	46
Figure 12	*Cross-section of part of a gasworks site, Gateshead*	50
Figure 13	*Massive foundations and cast iron gas pipes*	51
Figure 14	*Underground gas holder*	51
Figure 15	*Effect of self-heating on combustibles*	59
Figure 16	*Phases in landfill gas generation*	62
Figure 17	*Contamination seeping through brick lining of a tunnel*	65
Figure 18	*Flooding of tar over clean area due to careless demolition*	68
Figure 19	*Excavation of lime pit in former chemical works*	75
Figure 20	*Excavations covered by a car park*	77
Figure 21	*Foundations exposed in groundworks*	82
Figure 22	*Sealing boreholes through contaminated material*	84
Figure 23	*Boreholes for groundwater monitoring in contaminated fill*	85
Figure 24	*Scheme for analysis of solid and liquid samples*	94
Figure 25	*Groundwater samplers*	98
Figure 26	*Seeps of oil*	99
Figure 27	*Garages distorted by settlement of fill*	106
Figure 28	*Fifty-year self-weight creep settlement predicted from 6-month monitoring*	109
Figure 29	*Fifty-year building settlement predicted from 6-month load monitoring: shallow fill*	111
Figure 30	*Loading test on fill*	111
Figure 31	*Loading test on a pad footing*	112
Figure 32	*Stone columns in demolition debris fill*	133
Figure 33	*Dummy footing load test*	136
Figure 34	*Dynamic compaction*	137
Figure 35	*Evolution of cover layers in UK practice*	140
Figure 36	*Pumping water and tar from an underground gas holder*	141
Figure 37	*Basics of cover layer reclamation*	144
Figure 38	*Tar-covered bricks used as fill in an underground tank*	146
Figure 39	*Settlement of a house on a raft*	161
Figure 40	*Relative stiffness of raft, bending moment and differential settlement*	163
Figure 41	*Cross-section of floor slab showing natural ventilation spaces and louvres*	166
Figure 42	*Vents in external walls of houses*	167
Figure 43	*Louvres for underfloor ventilation*	167
Figure 44	*Monitoring for methane in permanent gas and water observation well*	175
Figure 45	*Tar tank below ground*	184
Figure 46	*Gun barrels found in Woolwich Royal Arsenal development*	188
Figure A1	*Coal gas production process*	210
Figure A2	*Refuse dated at 1965 and still capable of producing methane*	221
Figure A3	*Sampling landfill gas*	222
Figure A4	*Gas-sampling borehole*	223
Figure A5	*Recharge of landfill gas borehole after purging with nitrogen*	224

List of abbreviations

ADAS   Agricultural Development Advisory Service
AERE   Atomic Energy Research Establishment
ANC   Acid neutralising capacity
BOD   Biochemical oxygen demand
BRE   Building Research Establishment
CBR   California bearing ratio
CDEP   Central Directorate on Environmental Protection
COD   Chemical oxygen demand
CPT   Cone penetration test
CUR   Commissie voor Uitvoering van Research
C_v calorific value
DLG   Derelict Land Grant (see Appendix 1)
DoE   Department of Environment
DPT   Dynamic penetration test
DTp   Department of Transport
EC   European Community
EDTA   Ethene diamine tetra-acetic (ethanoic) acid
FRS   Fire Research Station
GLC   Greater London Council (formerly)
GRC   Glass-reinforced cement
HSE   Health and Safety Executive
ICRCL   Interdepartmental Committee on Redevelopment of Contaminated Land
LEL   Lower explosive limit
LPG   Liquefied petroleum gas
NATO   North Atlantic Treaty Organisation
NHBC   National House-Building Council
NRA   National Rivers Authority
NRPB   National Radiological Protection Board
OECD   Organisation for Economic Co-operation and Development
PAH   Polynuclear aromatic hydrocarbon
PCB   Polychlorinated biphenyl
pfa   Pulverised fuel ash and cycloned fly ash
ppm   Parts per million
PVC   Polyvinyl chloride
RPM   Reinforced plastics matrix
SPT   Standard Penetration Test
TOC   Total organic carbon
UDG   Urban Development Grant (see Appendix 1)
USEPA   United States Environmental Protection Agency
WRc   Water Research Centre

Glossary

Aerobic Pertaining to conditions where free oxygen is present.

Air-flush drilling Making a borehole by rotary drilling using compressed air to remove cuttings from the borehole.

Anaerobic Pertaining to conditions where there is no free oxygen.

Anemometer Instrument for measuring gas flow velocities.

California Bearing Ratio (CBR) A parameter used to determine road pavement thickness, it is the ratio (as a percentage) of the loads required on a plunger to penetrate a given amount into the tested material and a standard crushed rock.

Carcinogen Any cancer-producing substance.

Collapse settlement See Section 11.3.4.

Co-disposal Disposal of domestic refuse and commercial or industrial wastes in the same tip.

Compression As used in this publication, the vertical movement of a point in a material (usually at its surface) expressed as a fraction of the depth of compressible material below that point.

Consolidation The change in volume of a soil occurring under an applied stress as a result of dissipation of excess pore pressures.

Depth of fill In this publication, depths of fill are described as follows:
    shallow — less than 4 m
    medium — 4 to 15 m
    deep   — more than 15 m

Downdrag Downward frictional load on pile applied by surrounding stratum or strata settling relative to pile.

Drawdown The depression of the water table by pumping when extraction exceeds natural replenishment in the ground; also referred to as cone of depression when pumping from a single well.

Dust Particulate material which is or has been airborne and which passes a 200-mesh BS test sieve (0.076 mm).

Dynamic compaction See Section 18.5.

Fill site See Sections 1.3 and 11.1.

Flash point Temperature above which a flammable liquid starts to emit ignitable vapour.

Green field site A site which has not been developed or used for any commercial or industrial purpose and which therefore should be uncontaminated. It may include land used for grazing, arable or horticultural crops but not for intensive farming.

Gully erosion Phase of accelerated soil erosion in which thin layers of soil are removed with the formation of rills or gullies.

High wall The vertical or near-vertical working face (or abandoned working face) in a quarry or other mineral working.

Ignition temperature Temperature at which a gas may be ignited and combustion sustained.

Katharometer An instrument for the analysis of gases by the means of measurement of thermal conductivity.

Landfill Landfill is the term used in this publication for modern (i.e. after about 1975) commercial, industrial and domestic waste and refuse placed in a tip in a controlled manner with the intention of reclaiming the site for after use (*cf.* **Wastefill**).

Landfill gas The mixture of gases (see Section 12.11) generated from biodegradable material deposited in a landfill or wastefill. Its origin is not restricted to landfill as defined above.

Leachate In the context of derelict land, the fluid collecting at the lower horizon of a permeable fill, comprising percolating precipitation, groundwater and fluids contained in or arising from the fill and substances leached from the fill by the percolating fluids.

Listed/non-listed contaminant hazards These terms have the meaning assigned in Section 15.7.4.

Pathogen Any disease-producing micro-organism or substance.

Phytotoxic Toxic to plants.

Rainfall excess Condition when precipitation exceeds surface interception and infiltration, and surface hollows are filled, so that a thin sheet of surplus surface water starts to move overland.

Redox potential Measure of the potential reactivity of a substance in the presence of oxidising or reducing agents.

Sheet erosion Phase of accelerated soil erosion in which thin layers of soil are removed without the formation of rills or gullies.

Special waste Waste material so defined by regulations made under Section 17 of the Control of Pollution (Special Waste) Regulations 1980 (SI 1980, No. 1709) issued under the Control of Pollution Act 1974. Previously called 'notifiable waste' under the Deposit of Poisonous Waste Act 1972. Further explained in DoE Waste Management Paper 23, 1981. Such wastes may be deposited only at sites specially licensed to receive them.

Tailings Rejected portion of ore or that portion washed away in water concentration.

Target Anything exposed to a hazard, such as a person, animal or plant or a building or groundwater.

Vibroflotation See Section 18.4.

Wastefill Commercial, industrial and domestic waste and refuse placed in a tip (usually before about 1975). The term is used in this publication for a fill of these materials that would not be classed as landfill.

Water gas Mixture produced by decomposition of steam by incandescent carbon: typically, 45% carbon monoxide, 50% hydrogen, remainder carbon dioxide and nitrogen.

Part A

1. The present scene

The Department of the Environment currently accepts the definition of derelict land as: '... land so damaged by industrial or other exploitation as to be incapable of beneficial use without treatment ...' The Derelict Land Survey of 1988 recorded nearly 40 500 ha of such land in England alone, equivalent to a dereliction density of 0.33% of the total land area of the country. This was a net decrease from the 1982 figure of 45 700 ha (Figure 1), helped by the reclamation of over 14 000 ha with government grant aid in the intervening period. By the above definition, the net stock of derelict land appears to be diminishing slowly.

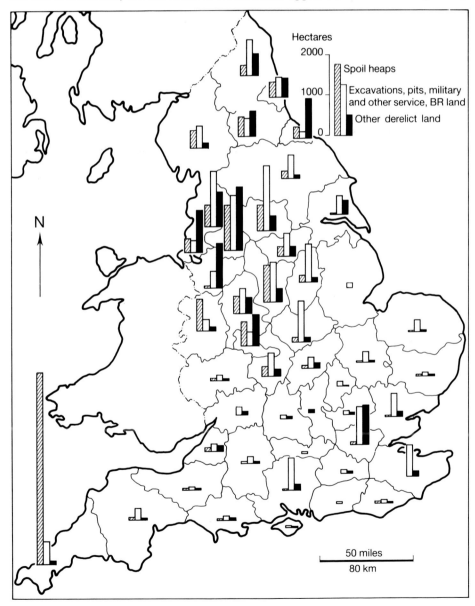

Figure 1 *Derelict land in England, 1982 (after Bridges, 1987, courtesy of Geography Department, Swansea University)*

That definition is only one of many proposed since 1946 (Bridges, 1987) and, in isolation, it is neither comprehensive nor entirely satisfactory, because land remaining in use may not have been recorded as derelict until abandoned. Apart from the resulting distortion in the figures, imprecision in terminology has led to a situation in which significant areas of the country could fail to qualify for legal recognition and financial grant aid. This publication is relevant not only to land already recorded as derelict as above but also to many other sites which, though factually derelict, have not yet been officially acknowledged as such.

Derelict sites are becoming increasingly available for reclamation, as old industries cease operation through obsolescence or insolvency and wastefill areas reach capacity. Experience over the past decade or so shows that most of this land can be brought back to beneficial use.

Such land may be cheap to buy, can often attract reclamation grants and, with efficient reclamation, its value is usually enhanced. It is often located in congested areas in cities where building space is urgently needed, and its development can help avoid the exploitation of greenfield sites. There is thus commercial and environmental advantage in the beneficial re-use of derelict land.

In the UK, legislative control of reclamation and development is contained in the Public Health Acts and the various Town and Country Planning Acts, which impose on local authorities the role of public safeguard. This lays special, and sometimes conflicting, responsibilities on local authorities, environmental health officers and engineers entrusted with the reclamation and building development of derelict land.

Derelict sites are of two basic types: fill sites and industrially contaminated sites, but many hazards are common to both.

1.1 Fill sites

Wastefill sites are not confined to specific industrial zones but are widespread wherever surface mineral extraction has left convenient holes in the ground. Quarries and open pits have long been used to dump domestic, building and industrial waste; these are often deep, to 30 m or more. Such fills are frequently contaminated, and hazardous materials may be present at any depth. Only those contaminants located in the upper 4–5 m below finished ground level are likely to be of direct concern to surface developments, but contaminants at all depths may be a hazard to aquifers and water resources.

Wastefills present many hazards: poor engineering properties, contamination and biodegradation effects, gas and leachate production, and underground burning. Even the older contaminated fills may still be chemically and biologically active, and leachate may be a threat to ground and surface water resources.

Many old wastefills in sand and gravel pits have been restored to agriculture. Under the limited definition given above, such sites might not be classified as derelict even though they would require extensive remedial treatment before they could be developed for building.

Recent opencast mining has left backfill sites of several hundred hectares each, which are generally free of biodegradable material or contamination. In the past these backfills, dumped by dragline and allowed to settle under their self-weight after rough levelling by bulldozer, were usually restored to agriculture. Although perhaps not strictly derelict sites by definition, these fills would require extensive densification treatment before they could be considered for building works. In a few special cases, where an opencast site had already been earmarked for building development or roadworks, a substantial upper thickness of the backfill was carefully densified by spreading in layers and compacting. On some other sites during the past decade or so, improvements in the backfilling methods (e.g. using scraper deposition) have also led to much-improved engineering properties, approaching the quality of specially compacted backfill (Knipe, 1979).

Usually, however, the most serious handicap to building development on fills is the potential for large and uneven settlements, requiring extensive remedial ground treatment and careful attention to foundation design. Gas generation may also rule out some wastefills for building works.

Figure 2 *Pear Tree Lane Colliery, Dudley, c.1890*

1.2 Industrially contaminated sites

Practically every town in the Black Country, the Potteries and elsewhere in the Midlands, the industrial North, Clydeside and South Wales and, to a lesser degree, south-east England has derelict areas of industrial contamination. Foreign, potentially harmful substances remain in the ground or accumulate by tipping, airborne or waterborne deposition. The contamination is often found to have migrated beyond its deposition zone. Some contamination dates from the Industrial Revolution or earlier, and often is the result of a succession of different industries using the same site (Figure 2).

In contrast to fill sites, the contamination of industrial sites is usually shallow, and the ground is often well compacted so that settlement may not be a problem. On the other hand, deep and massive foundations, buried tanks, services, etc. often remain as obstacles to development.

Some contaminants may react dangerously when disturbed or upon a change in the water table. Chemicals in the ground may thus pose a threat to the health and safety of human, animal and plant life and to the buildings, materials, foundations, services and water resources.

Contaminated land may also emit gases (usually to a less extent than fill sites) and may present a fire hazard (mainly from coal residues, refuse, and oil and solvent spillages).

1.3 Essentials and implicit responsibilities of development

Development of a derelict site is intended to produce a facility which shall be safe for future occupants or users, and without danger to third parties and the environment. Safety must be paramount above all other considerations, and no compromise can be admitted. A development which poses a threat to occupants or the environment is unlikely ever to be a viable proposition. On the other hand, there will always be some risk attendant on the activities of site investigation and reclamation. Stringent precautions are necessary to ensure that these risks can be faced without harm to workers on the site.

Other desirable features (such as developer's preferences, local needs, ease and economy of construction, maximum return on investment, etc.) may not always be compatible, either mutually or with the over-riding requirements of safety. On such issues some compromise may be necessary, and various options are open during the course of development.

There is increasing concern about the potential migration of liquid or gaseous contamination from a site, whether or not it is to be redeveloped. The responsibility for abatement can be assumed to exist notwithstanding any decisions on whether development is appropriate or not. This issue was highlighted in the House of Commons Environment Committee Report on Contaminated Land (HMSO, 1990).

2. Logical procedure and structure of this publication

The suitability of a site for a particular form of development will depend on the presence, or otherwise, of hazards likely to affect the end use created by the development. There may be a great number of hazards on a given site, but relatively few are likely to affect a particular end use, and not all of these will be of sufficient intensity to pose a threat. To save an unnecessary search for irrelevant hazards, logical procedure starts by proposing an end use, and investigates those hazards known to affect that end use and whose presence may be expected on the site in question.

This publication sets out the procedure for determining the suitability of a site for a particular end use and the necessary measures for the realisation of a suitable development. The text is structured to follow the sequence implicit in the logical procedure, and starts with an exposition of essential background information on derelict site hazards and associated problems and constraints.

Parts A and B both follow a similar logical sequence, thus:

PART A	COVERAGE	PART B
3. Technical problems and professional expertise	**Basic knowledge** Background information and guidance on hazards to be expected on various types of derelict site and their probable adverse effects on particular end uses. Hazards may call for multi-disciplinary expertise.	11. Engineering problems of fill 12. Hazards of contamination
4. Potential constraints	**Constraints** Site hazards result in constraints on freedom of action: statutory controls, safety requirements, time-consuming and costly working procedures.	13. Constraints to development
5. Investigation strategy	**Initial assessment** Site visits and study of site history give first indications of possible hazards likely to affect the proposed end use.	14. Initial assessment
	Site investigation Detailed investigation of the site provides quantitative data on hazard intensities.	15. Site investigation 16. Sampling techniques
6. Site appraisal	**Site appraisal** • Comparison of measured hazard levels with acceptable safe reference values. • Options open if hazards render desired end use unrealisable under existing site conditions: — change of end use or layout — site reclamation/protective design. • Decisions on definitive end use and exercise of options.	17. Detailed site appraisal

7. Remedial
   treatment
   of ground

Site reclamation
Remedial treatment to reduce adverse
effects of hazards, so as to bring
site towards a condition of
compatibility with the definitive
end use.

18. Remedial treatment
    of fill sites

19. Remedial treatment
    of contaminated sites

8. Design

Design measures
Protective design (foundations, services,
structures, etc.) is intended to bridge
incompatibility remaining between the
reclaimed site and definitive end use.

20. Design measures

9. Monitoring

Monitoring
A derelict site is a dynamic, evolving
system. Monitoring desirable.

21. Monitoring procedures

10. Contractual
    considerations

Contracts
Contracts for derelict land development
have to take hazards and constraints
into account.

22. Contractual guidance

Workers' safety
Protection of site workers calls for
special training, facilities, equipment
and working procedures.

23. Safety in site working

3. Technical problems and professional expertise

The fundamental question, to be answered before building development of a derelict site can be contemplated, can be framed as:

> Is the condition of the site compatible with the proposed end use and, if not, what measures are necessary (or possible) to reduce the incompatibility to acceptable proportions?

The full resolution of this question is only reached by a procedure of investigation, appraisal and engineering design. This procedure is outlined in later Sections of Part A and detailed in Part B.

To begin the procedure requires an understanding of two interacting relationships: (1) sites and hazards; (2) hazards and end uses.

3.1 Sites and hazards

Derelict land exhibits hazards, usually the result of human activity, which are not normally encountered on greenfield sites. Table 1 is a summary of the more common hazards.

Table 1 Potential hazards of derelict sites

Physical
Heterogeneous nature of the ground
Poor engineering properties (affecting bearing capacity and settlement)
Obstructions (e.g. old foundations, pits, tanks, services, massive fill materials)

Chemical/biochemical
Health and environmental hazards (from substances affecting human, animal and plant life—including carcinogens and, occasionally, radioactivity)
Biodegradation effects (e.g. decomposition products and underground voids)
Combustibility (of materials above and below ground)
Aggressive attack, and corrosion of building materials
Gas emissions (flammable, toxic or asphyxiant)
Groundwater pollution and leachate production (affecting water resources)

Physico-chemical
Expansive reactions of some slags, mining spoils, etc. under foundations and services

Poor engineering properties, settlement and biodegradation problems are usually associated with fills. With those exceptions, the other hazards in Table 1 may be found on both fill and contaminated sites, and no distinction is made at this stage between the two. In Part B, further guidance is given about the relationship between hazards and types of site.

Section 11, *Engineering problems of fill*, and Section 12, *Hazards of contamination*, are intended to give the engineer the technical background to assist in a properly informed appraisal of a derelict site and the formulation of economic technical solutions. These two sections are separated purely for convenience of exposition; many hazards are common to both. They consider each of the various hazards in isolation—it is for the engineer to assemble the appropriate combination of information to match the site under consideration.

3.2 Hazards and end uses

The severity of the problem posed by a particular hazard will depend on the sensitivity of the various targets exposed to it, in consequence of the proposed end use. Table 2 is a useful preliminary guide to the complex of relationships for the more common types of derelict site, hazard and end use (or target). Note that Table 2 is printed inside the back cover of this publication so that it can be referred to when reading other sections. The upper part of the table indicates the likelihood (*) of hazards being a problem on particular types of site, while the lower part shows the sensitivity (O) of various targets to these hazards.

Table 2 shows, for example, that old industrial sites are very likely (∗∗∗) to contain toxic organic substances, migratory contaminants, oils and tars, and heavy metals, all of which could have serious implications for kitchen gardens (highly sensitive target, ○○○). Consequently, severe technical problems would have to be overcome before an old industrial site could be dedicated to an end use involving houses with kitchen gardens. Likewise, untreated domestic fill is seen to impose serious constraints on building works by reasons of settlement, gas emission, combustion potential, aggressive chemical attack, etc.

Even sites of the same type differ in the intensity of their characteristic hazards, and an indicative table of this sort cannot pretend to be comprehensive. However, it gives warning of the hazards likely to affect a given site and, until experience has been gained by other means, it may serve as a convenient starting point for preliminary discussions on land use.

3.3 Professional expertise

It is unlikely that any single professional discipline will possess sufficient expertise to confront unaided the problems raised by the range of hazards on most derelict sites. The engineer should at all times be alert to the need for, and the advantages of, a multi-disciplinary approach—both during the progress of site investigation and appraisal as well as in the subsequent reclamation works. Specialists whose advice and collaboration may be needed, and the aspects on which they should be suitably experienced include:

- Environmental scientist   Fill and contamination hazards in general; protection of environment, water resources, etc.; control of leachate, gas, combustion; planning consent procedures
- Chemist   Detailed investigation, identification, measurement and treatment of specific contaminants, leachates, gases
- Geotechnical engineer   Behaviour of fill; ground improvement; foundation design
- Hydrogeologist   Groundwater and leachate regimes
- Corrosion scientist   Corrosion of works below ground
- Botanist   Condition of existing vegetation, new planting and ground cover
- Safety specialist   Safety of site workers and third parties
- Demolition specialist   Site clearance and safe dismantling of existing works and structures
- Industrial specialist(s)   Operation of industries previously on site.

From the outset the engineer should ensure that the developer understands the need for such expertise, and is willing to include the requisite expenditure in the budget.

4. Potential constraints

The hazards of a derelict site impose constraints on freedom of action, not only on the contractor in site operations but also, more fundamentally, on the choice of a suitable development as well as on the statutory authority in giving planning consent. Protection of the environment may dictate what form of reclamation is permissible and the means of disposal of dangerous material. Safety of the works and future occupants may require a measure of over-design, while the protection of site workers will demand time-consuming and costly safety procedures and facilities.

It is convenient to consider constraints under five related headings:

1. Technical and environmental
2. Statutory
3. Safety
4. Time
5. Cost.

4.1 Technical and environmental

Experience over the past decade or so has shown that there are very few derelict sites which cannot be brought back into beneficial use. Inherent hazards may, however, impose constraints on the choice of the form of development; these will become evident on matching the proposed end use to the hazards (e.g. as in Table 2).

Constraints may necessitate revision of the development plan, such as a change in the layout to locate sensitive land uses in the less hazardous areas. Alternatively, in the case of severe hazards, a change to a less sensitive end use may be advisable. For example, heavy metal contamination in the immediate subsoil could demand extensive remedial measures before the site could be dedicated to housing with gardens, but treatment would be unnecessary if the site were to be used for industrial development. Before opting for a low-sensitivity end use, a developer (and the local authority) should be alerted to the possibility of a reversion to more sensitive end use in the future. However, it is generally accepted that site treatment to suit all conceivable future uses is not justified. The best policy is to record carefully what is known about the site history, the ground and the contamination, and how the site has been reclaimed.

The condition of some abandoned derelict sites may demand immediate action for public safety, such as restriction of access, even while the site remains undisturbed. Once a development starts, even a 'safe' site becomes potentially hazardous and requires detailed precautions.

Landfill gassing sites may be unsuited for building development. Good practice would normally counsel against building above active gas-producing land and adjacent areas which might be affected by lateral migration of gas. The putrescible organic source of the gas may also constitute an underground fire risk.

On a fill site, settlement potential may impose severe constraints on building development, and extensive ground improvement and careful foundation design may be required before building works can be realised. In some cases a significant technical constraint to building may be the presence of aggressive chemicals liable to attack construction materials. On many old industrial sites coal residues and other potentially combustible materials may also require remedial measures.

In addition to the direct effects of soil contamination, derelict sites can also produce contaminated leachate, which may demand special measures for the protection of water resources and the environment in general (Figure 3).

4.2 Statutory

Considerable legal and statutory constraints apply to derelict land development, for the protection of the environment and prevention of nuisance, as well as for the protection of

Figure 3 *Layers of old contaminants (photograph by courtesy of GLC)*

health and safety of site workers, third parties and future occupants. In addition to control under the Public Health Acts and the Building Regulations, a developer will be bound by legislation on planning, waste disposal, pollution control, etc.

4.3 Safety

Safety of the works, future occupants, site workers, third parties and the environment must be considered at all stages. Design of the works must allow for their safe realisation on site. Acceptable safety margins on hazardous sites may require a degree of over-design compared to a greenfield. The safety of site personnel is the responsibility of the contractor, and will require time-consuming working procedures and the provision of special facilities, clothing and equipment (Section 23). Safety of the environment and natural resources is now of national concern, and EC regulations are likely to become increasingly stringent.

4.4 Time

Constraints arise not only from the necessarily lengthy appraisal and reclamation stages buy also frequently from the time required to obtain planning permission and other statutory approvals. Additionally, the developer may have difficulty in obtaining funds until all technical and statutory constraints have been determined.

4.5 Cost

Derelict land can be cheap to buy, but remedial treatment to bring it to a suitable condition for development can be costly. In addition, the necessary design measures for the development may be dearer than for a greenfield site. Nevertheless, building on derelict land is an attractive option for developers. Of the 14 000 ha reclaimed between 1982 and 1988, some 4100 ha were reclaimed without grant by agencies other than local authorities, and about 1300 ha by the private sector and Urban Development Corporations with grant aid (DoE, 1989).

Even so, cost constraints can be sufficient to entail revisions to a development scheme before it can be judged economically viable.

4.6 Consequence of constraints

In consequence of the many constraints operating on a derelict site, it is inadvisable for a developer to undertake a financial commitment until sufficient information on the hazards and other constraints has been obtained for a full appraisal by the engineer.

The content of the essential information, and the manner of obtaining it, is outlined in Section 5. A development undertaken without such information risks bringing to light unsuspected hazards during its realisation or, worse, after its completion. The cost of remedies as unplanned emergency measures will almost certainly outweigh the avoided cost of adequate prior investigations. Hazards coming to light after completion of the development may even be grounds for legal action.

5. Investigation strategy

5.1 Scope

Before attempting to design the works or plan the reclamation of a derelict site, or even before advising on the feasibility of a proposed development, the engineer has to make at least a preliminary appraisal of the site and of the consequences for a chosen end use. The information on potential hazards and constraints affecting the site, which is needed for that appraisal, may include some or all of the following:

- Settlement characteristics and buried obstructions
- Health and safety hazards in the long and short terms
- Combustibility (or actual fire) at surface and underground
- Aggressive attack on building materials
- Biodegradation effects
- Gas generation, migration and emission
- Groundwater pollution; leachate generation, migration and regime, and threat to surface and groundwater resources and run-off
- Expansive reactions in the ground
- Immediate environmental or safety hazards calling for emergency action
- Statutory and other constraints.

For the definitive design of the works a full appraisal is essential, for which the above information has to be in thorough analytical detail; simple qualitative observations are not enough. For example, if gas is present the information provided by a simple gas detector will not suffice. Instead, controlled measurements should be taken to establish the emission rate in the ground and, consequently, the rate of accumulation in enclosed spaces.

At this stage no distinction is made between fill and contaminated sites, and much of the foregoing will apply to either. Part B gives the background necessary to permit the formulation of an investigation programme appropriate to a given site.

Current practice is to obtain the requisite information in two distinct phases:

> *Phase 1* Initial assessment of site (Sections 5.2 and 14)
> *Phase 2* Detailed site investigation (Sections 5.3 and 15)

Only rarely will the information obtained in Phase 1 be sufficient by itself for a full site appraisal without recourse to the more costly and complex operations of the detailed investigation.

The investigation of contamination should conform to the recommendations of the Draft for Development, DD175, *Code of Practice for the identification of contaminated land and its investigation* (BSI, 1988). In addition, BS5930 *Code of Practice for site investigation* (BSI, 1981) and CIRIA Special Publication 25, *Site Investigation Manual* (Weltman and Head, 1983), although intended for greenfield sites, contain much that is relevant to derelict sites.

5.2 Phase 1: initial assessment of site

Scope. This is intended as a site-specific operation, preferably before purchase, for obtaining qualitative information by simple means, involving as few people as possible and with minimum exposure to the still-unknown site hazards. (It is termed 'preliminary investigation' in DD175.)

Its primary purpose is to identify the nature and general location of contamination or other hazards likely to affect a proposed development. Under favourable circumstances it may also give preliminary indications as to the suitability of the development, its probable cost, and advisable modifications, as well as possible constraints. More importantly, it should provide the basis for a detailed programme of site investigation (Phase 2), including also estimates of cost and resources required, and measures for containment of contamination and the safety of the more numerous personnel likely to be involved.

Although it is conducted by relatively simple means, the importance of initial assessment should not be underestimated. The longer a hazard remains undetected, the greater is the risk of needless expenditure on operations subsequently proved unrealisable once the hazard comes to light. Care taken during this early phase of investigation can materially reduce this risk.

Initial assessment comprises two distinct operations: desk study and site reconnaissance.

Desk study. This involves the acquisition of sufficient documentary and other information to define the industrial processes and other activities likely to have contaminated, damaged, obstructed or otherwise degraded the site. To that end, a thorough review is required of all available historical and other evidence relating to previous usage of the site and the geological and physical nature of the site and its environs, as well as information as to statutory and other constraints.

Site reconnaissance. This involves visits to the site and its environs, and provides visual evidence of hazards and obstacles to development. It should include locating, and sampling from, existing boreholes, springs and surface waters. In cases where contaminant migration may have occurred (e.g. gas, leachate, oil and tar impregnations, etc.) the reconnaissance should extend to adjacent areas beyond the confines of the site.

In order to avoid risks to personnel from unknown hazards the site reconnaissance should be delayed, if practicable, until the desk study has yielded sufficient information for appropriate precautions to be taken.

5.3 Phase 2: detailed site investigation

Scope. This is a detailed analytical study intended to provide quantitative information on hazards, with the object of a full site appraisal. It should provide data in sufficient detail to permit definitive decisions on:

- Financial and technical options open to the development
- Planning of remedial treatments
- Design and realisation of the works
- Safety measures, both long and short term
- Need for long-term monitoring.

It involves a campaign of accurate survey, sampling, *in-situ* and laboratory testing, analysis and monitoring, and requires experienced personnel, as well as experts in appropriate disciplines. It also requires adequate safety precautions for the investigation workers (Figure 4).

Given the heterogeneous nature of derelict land, it is rarely economic to examine the whole site in fine detail. In practice, available data should be used to direct the investigation to specific areas. A preliminary examination over the whole site should be followed by closer investigation of selected areas, designed to confirm areas of suspected high- or low-hazard intensity.

Subsequent operations such as piling, excavation, foundations, drainage, etc. expose much more of the ground in depth than is practicable during site investigation. The engineer has to be alert to the possibility of unearthing new hazards during such works, and should maintain access to expertise qualified to identify and deal with them. Previously unsuspected hazards may enforce revisions to the project. This means that it is usually not possible to define at the outset the extent of the reclamation works. The developer should be warned of this and advised of the need to adopt a flexible approach.

Conventional investigation techniques for all types of site. Certain basic routines are common to the investigation of all types of site, whether greenfield or derelict, and include:

- Survey and setting out of sampling points
- Execution of trial pits, trenches and/or boreholes
- Sample extraction, handling, containment, transport
- Sample identification, data logging and testing

Figure 4 *Underground combustion (photograph by courtesy of Allott and Lomax)*

- Groundwater sampling
- *In-situ* geotechnical tests
- Monitoring of groundwater levels.

Additional special investigations for derelict sites. The following operations, *inter alia*, may be required for obtaining the special information called for in Section 5.1:

- Determination of extent, depth and cross-section of deposit, and nature and composition of deposited materials
- Determination of type, age, degree of saturation, compaction and permeability of deposited materials, and principal and perched water tables
- Monitoring of ground levels and load tests over extended periods
- Recording surface deposits and other indicators
- Sampling deposited materials in depth for chemical and biochemical tests
- Laboratory tests either on- or off-site for chemical/biochemical identification of samples (relevant to potential health and contamination hazards, combustibility, aggressivity, expansivity, biodegradability, etc.)
- Surface and underground temperature measurement
- Gas sampling, analysis and flow determination in boreholes
- Water sampling for chemical testing; monitoring of groundwater composition and movement; pumping drawdown tests.

5.4 Safety in site investigation

The hazards present on derelict land can pose a threat to site workers, and the safety principles outlined in Section 23 apply to all phases of the works. From the first entry on site, investigation workers will be subject to hazards, many still undiscovered, but the investigation contract value is rarely sufficient to support the sophisticated safety measures of the reclamation and construction contracts. Nevertheless, appropriate safety measures must be taken and investigation works should be pre-planned to permit rapid and effective working and reduce to an acceptable level all exposure to *in-situ* hazards.

6. Site appraisal

6.1 Scope

Appraisal is concerned with determining the technical and economic suitability of a site for a specified end use, and with measures to achieve suitability. The site investigation should reveal the presence, or otherwise, of those hazards likely to affect an end use. As results come to hand, a continuing process of evaluation should start by comparing measured hazard intensities against 'safe' acceptance criteria in order to appraise the effects of discovered hazards on the end use.

If, on appraisal, discovered hazards show the site to be incompatible with a proposed end use, consideration should be given to the options available (outlined below in Section 6.3). This may entail a decision to change the end use or the layout of the projected development. Sometimes it may be feasible to retain the desired development scheme, with appropriate reclamation of the site and design of the works. The exercise of these options will, however, necessitate a re-evaluation of such other hazards as may have become relevant under the changed conditions. If the options bring into play hitherto unexplored hazards, further site investigation will also be required.

In this manner, by continuing process of trial and error, appraisal arrives at definitive decisions on the most suitable end use, remedial treatment and design measures for the particular site.

6.2 Acceptance criteria for hazard levels

The first part of the original question posed in Section 3, concerning the compatibility of the site and the proposed end use, can now be reframed as a subsidiary question in closer detail:

Does the measured intensity of a discovered hazard exceed acceptable values sufficiently to constitute a threat to targets exposed by the intended end use?

The resolution of this question presupposes an established set of criteria of acceptability as a yardstick against which to judge the measured hazard intensities. Current practice recognises four distinct types of acceptance criteria:

1. Hazard unacceptable at any level (Type A)
2. Acceptance level defined by statutory authority (Type B)
3. Acceptance level defined by current medical practice (Type C)
4. Acceptance level defined by standard technical practice (Type D).

At the time of writing the available acceptance criteria are far from comprehensive, and their practical application calls for experience and careful judgement. Section 17 outlines the available criteria and their application to various types of dereliction hazard.

6.3 Options for action

If the appraisal shows the condition of the site to be unsuitable for a proposed end use, three further steps—*planning, remedial works, design*—provide a series of options for reducing the incompatibility to acceptable dimensions:

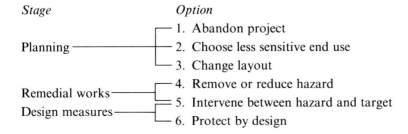

Stage	Option
Planning	1. Abandon project
	2. Choose less sensitive end use
	3. Change layout
Remedial works	4. Remove or reduce hazard
Design measures	5. Intervene between hazard and target
	6. Protect by design

If the planning options (1, 2 or 3) are unacceptable or insufficient by themselves, remedial works (options 4 and 5) may be exploited to increase the suitability of the site, while design measures (options 5 and 6) are intended to bridge any remaining incompatibility. Section 17.2 deals in depth with the exercise of these options in practice.

Remedial treatments and design measures are studied together so as to arrive at the optimum combination to match the definitive end use to the site. If the first result is technically or economically unattractive, changes are made through the exercise of the various available options. The process is repeated until the best combination of site/remedy/design/end use is found.

7. Remedial treatment of ground

Sections 18 and 19 in Part B describe remedial measures to reclaim a site, that is, to bring the site towards a condition compatible with the proposed end use. Two options are available:

Option 4   Removal or reduction of hazard
Option 5   Intervention between hazard and target.

7.1 Fill sites

Remedial treatments are concerned principally with the reduction of potential differential settlement to be suffered by a development bearing directly on the fill. This is achieved by improving the load-carrying properties of the fill prior to development, generally through increasing its density by one of the following techniques (Section 18):

- Excavation and substitution with sound material, or replacement in layers and re-compaction (type 4 option)
- Compaction under temporary pre-load (type 4 option)
- *In-situ* treatment by specialist processes (type 4 option) (Figure 5).

With all these expedients some residual settlement must be accepted. They are, therefore, normally used in conjunction with raft-type foundations (design option 6).

Figure 5 *Vibro-replacement (stone columns) in fill (photograph by courtesy of Cementation Piling and Foundations Ltd)*

The alternative to remedial treatment, aimed at reducing settlement effects to insignificance by positive support through piling down to an underlying competent stratum, is dealt with separately as a type 6 design option in Section 20.

7.2 Contaminated sites

In the UK two methods are currently in vogue for the treatment of contaminated sites (Section 19):

1. Reduce accessibility of contaminant under designed cover layer, and/or sometimes within barrier walls (type 5 option). Complete isolation of contaminant is not usually practicable.
2. Excavate the contaminated ground, remove to an approved tip and replace with clean material (type 4 option).

Other methods involving *in-situ* treatment or clean-up of soils on- or off-site are at present the subject of intensive research internationally, as are methods for cleaning contaminated groundwater. Many are based on well-established practices in other disciplines, and show promise for the future, but very few are at the stage of practical application in the UK, although their use is growing overseas (Smith, 1985 and 1987b).

7.3 Special considerations

Remedial treatment of derelict land, whether filled or contaminated, may also involve elimination of, or protection against, some or all of the following special potential hazards:

- Combustibility at surface or underground
- Aggressive attack and corrosion
- Expansion of deposited materials
- Gas generation and emission
- Groundwater contamination; leachate formation and migration.

For convenience, these are dealt with in Section 19.

7.4 Necessary and routine works common to all derelict sites

Upon acquisition of a derelict site a developer also acquires a responsibility to third parties to maintain, or relinquish, the site in a safe condition and not to create nuisance or damage to the environment.

Site to be abandoned. Should the developer decide to abandon the project (or otherwise to delay development) without reclamation, certain measures will be immediately obligatory for public safety, and may be ordered by the statutory authority. Depending on the conditions, necessary actions may include:

- Fence the site and erect warning notices
- Remove or cover visible surface contamination
- Divert services clear of the site, or ensure that any future excavation is strictly supervised
- Sink and monitor boreholes for continuing assessment of leachate risk to aquifer, and gassing risk to adjacent land
- Divert, or otherwise treat, streams, ponds, etc. to the extent necessary or practicable to avoid harm to local water resources
- Institute measures against risk of combustion, dust, etc.

In addition, the engineer will be required to prepare, and obtain statutory approval for, contingency plans against deterioration of site conditions. The engineer should also inform the local authority, water authority, gas and electricity boards, police and fire services, and provide the names of persons to be contacted in an emergency.

Site to be developed. In this case there are certain additional, essential routine works to be executed, irrespective of end use, whenever the site conditions are judged sufficiently serious:

- Clear site: including demolition, re-routing of services, removal of visible contamination, suppression of dust, etc.
- Regrade surface to control run-off
- Protect water mains
- Control groundwater migration
- Control gas migration.

8. Design measures

Remedial treatment (options 4 and 5) is intended to bring a derelict site towards a state of compatibility with a desired end use. It is thereafter the function of design (options 5 and 6) to bridge any incompatibility still remaining. For example, densification of fill will not eliminate settlement entirely but, in the majority of cases, can reduce it to acceptable proportions. Foundation and superstructure then have to be designed to protect the superstructure itself against the residual potential of the ground for differential settlement (option 6). Thorough remedial treatment of the ground is usually more effective than expedients relying solely on defensive design measures. In practice, remedial ground treatments and design measures are studied together, so as to arrive at the best combination to match the site to the definitive end use in an economic solution.

Table 3 lists various applications of the design measures discussed in Section 20.

Table 3 Application of design measures

Targets	Hazards
Foundations	Residual settlement
	Combustion
	Gas emission
	Corrosive attack
	Migrating contaminant
Buried services	Settlement
	Combustion
	Corrosive attack
	Migrating contaminant
Superstructures	Differential settlement
	Gas accumulation

9. Monitoring

Individual site investigation tests may yield information on the temporary state and intensity of a given hazard, but can give little indication of likely future changes. A derelict site is a dynamic system in which values may change over an extended period. Therefore early monitoring is desirable before reclamation and construction for the proper appraisal and design of a development. Equally, continued monitoring during and after construction is desirable for ensuring the continuing safety of the development and the local environment. Section 15.3 outlines aspects requiring monitoring and Section 21 the requisite procedures.

Monitoring before development provides information essential for the planning of remedial treatments and design measures. Thus the most dependable prediction of the settlement likely to be suffered by a building on fill is derived from accurate monitoring of load tests and ground levels. Likewise, gas emission-rates may vary seasonally, or may decrease with depletion of biodegradable material, or increase upon disturbance of a site. Since gas may well be the most serious constraint against a building development, monitoring of this potential hazard on a suspect site should start as soon as possible.

Continued monitoring during and after development serves to sound an advance warning of any change likely to constitute a danger, so as to permit the timely activation of countermeasures. Some long-term monitoring can be performed by automatic data-logging and control. In other cases, such as monitoring rates of contaminant migration, hand sampling may be needed at intervals.

10. Contractual considerations

There are two essential contractual differences, albeit of degree only, which distinguish the development of a derelict site from that of a greenfield:

- Constraints on the Contractor's freedom of operation
- Uncertainty as to the extent and intensity of hazards.

Constraints (already briefly outlined in Section 4, and dealt with in detail later in Section 13) have the affect of increasing the complexity of, and time required for, individual site operations. Moreover, many reclamation works involve highly specialised and continually evolving technologies. The satisfactory operation of these technologies requires considerable equipment, expertise and, above all, painstaking diligence and attention to detail on the part of the Contractor. This means that the unit costs of work items will almost inevitably be higher.

More seriously, it also means that, unless care is taken in specifying, it may not be possible to ascertain precisely what each tenderer has included in his price. This could be to the disadvantage of the more experienced and conscientious contractor. Therefore, instead of merely prescribing the performance required of the finished works, the Engineer should define very carefully the work to be done, and the constraints (safety of personnel, environment, etc.) attendant on its execution.

Uncertainty in the main (reclamation) contract arises from the impossibility of demonstrating that the site investigation has revealed the full extent and intensity of the hazards, and thus the extent of the work required. Even with the best information, there remains always the risk of unearthing further hazards during the course of the reclamation. To require the main Contractor to bear the consequences of such a risk would provoke unrealistically high tender prices. The effect of the foregoing is that it is usually not possible to define at the outset the extent of the Contract itself.

These two considerations have a profound influence on the type of contract most suitable for a derelict site, and on the specialised knowledge and care required of the Engineer in preparing contract documents and specifications. Detailed guidance on meeting these requirements is given in Section 22.

Part B

11. Engineering problems of fill

This section deals exclusively with the engineering problems of fill considered as a load-supporting medium. These derive from the physical properties affecting its bearing capacity and settlement characteristics, as well as from the chemical properties of certain expansive fills liable to cause disruption of foundations, etc.

Other chemical/biochemical characteristics of fill (as listed in Section 11.2) are described in Section 12 under the hazards of contaminated sites. In practice, derelict sites usually exhibit a combination of the various hazards considered in Sections 11 and 12.

11.1 Type and content of fill sites

Much of the surface of Britain is not the original soil as laid down by natural geological processes. Fill sites may contain any combination of poorly compacted soil and rock, a great variety of industrial, mining or metallurgical waste and scrap, and all types of domestic and commercial refuse, as well as brick, concrete or rubble from demolition work. The material may range in size from fine cohesive deposits, requiring time for consolidation, to massive blocks, causing problems for pile driving.

In some cases the growth of townships or similar developments conceals evidence of earlier filling; detailed site investigation and perusal of old records may be necessary to reveal their existence. Not all fill sites are derelict.

Quarries, sand and clay pits, gravel pits, etc. have traditionally formed convenient dumping grounds for a large range of wastes. Before the passing of the Control of Pollution Act 1974 wastefilling was usually executed with little thought for the future use of the site. There was scant restriction on the types of harmful material allowed in normal waste fills or their location or manner of deposition, and detailed records may not exist. Present-day records also vary greatly in quality and there is no statutory requirement for their preservation after completion of filling.

Domestic and industrial wastes have commonly been used for eliminating unwanted natural hollows, raising levels of flat land, and other landscape alterations. Many areas of past industrial activity are also found to be covered with extraneous material, in layers or haphazardly tipped.

Extensive opencast mining activities have left large areas (up to several hundred hectares) of overburden severely disturbed and usually in loose condition. In British opencast coal fields the ratio of overburden to mineral is generally high, so that mining has produced large areas of deep, inert, loose backfill (generally up to 30–40 m deep, but sometimes over 100 m). The backfill may be partially compacted, but rarely to a high or uniform standard throughout. Features such as washery lagoons and overburden heaps can cause anomalies in the settlement of the backfill. These sites are usually restored to agriculture.

Deposition of waste under water, producing fine, uniform cohesive deposits in lagoons, has long been the practice in the disposal of mine tailings, pulverised fuel ash (pfa) from power stations, and in the sedimentation of sewage sludge. Consolidation may be very slow, and water contents may be very high for many years after placement. Superficially dry (dewatered) materials may overlie deep deposits of wet material, giving a false impression of the 'engineering' capability of the site. This is typified by the crust forming over pfa lagoons.

There is thus a very wide range of fill sites and of filling materials. Charles (1984) has given a comprehensive table of commonly encountered fills, with brief accounts of their more important characteristics.

11.2 Problems caused by fill site hazards

Of the hazards listed in Table 1 those most likely to be encountered on fill sites include:

Physical
- Poor bearing capacity, settlement and obstructions

Chemical/biochemical
- Health and environmental hazards
- Biodegradation effects
- Combustibility
- Aggressive attack
- Gas emission
- Leachate production

Physico-chemical
- Expansive reactions

The severity of these hazards is variously related to the age, nature and condition of the fill.

Physical problems. Generally it is the settlement potential which gives rise to the most serious problems for building development on fill. Even where the appropriate remedial measures are taken, the geotechnical properties of the fill may often be unsuitable for all but simple light-weight buildings.

Particular considerations apply to domestic refuse. Figure 6 shows that, since 1940, the reduction in coal fires has led to a notable drop in the ash content of domestic refuse, with a corresponding increase in putrescible matter, plastics and paper. This has had the effect of lowering the bulk density and impairing engineering properties. Not only does the decay of putrescible materials produce flammable/explosive gases, but some putrescible materials are themselves intrinsically combustible. From the combustion of refuse, plastics, paper, etc. derive further hazards of fire, toxic and asphyxiant gases and the creation of voids in the ground. The effect is to the further detriment of the already poor engineering properties of the fill.

Other physical problems occur in fills, some of minor effect such as the stability of the weak and variable materials at the edges of the fill and some of major significance such as downdrag on piles caused by settling fill.

The technical problems posed by obstructions are outlined in Section 12.3.

Chemical problems. In comparison with industrially contaminated sites, fill site contamination may be too deep for economic removal, while gassing and combustibility risks may be sufficiently intractable to enforce the abandonment of a building project. Aggressive ground conditions on a fill site may be less intense than on an industrially contaminated site, due to lower concentrations of contaminants, but their effect may be aggravated by the larger area over which the groundwater can migrate. Perched water tables may be a local problem within the body of a fill, but leachate formation is frequently a serious concern of the local water authority, which has the power to impose severe restrictions on a developer. Technical problems may arise from the need to control and dispose of leachate.

11.3 Engineering behaviour of fill

11.3.1 Bearing capacity and settlement

It may not be feasible to convert the entire bulk of a fill to the engineering quality of a greenfield site. This is largely due to the considerable volume and depth of the fill itself, its intrinsically poor engineering qualities and, coincidentally, its contaminant content.

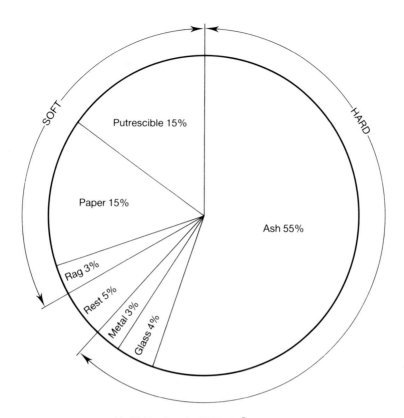

(a) 1940 – Density 270 kg/m^3 as collected

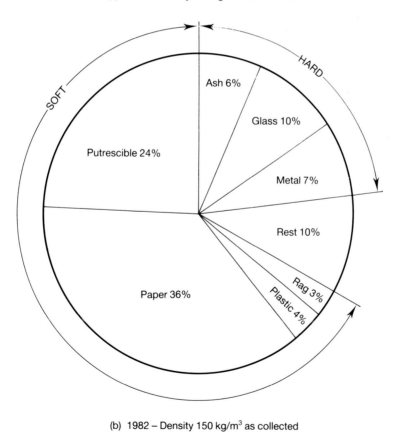

(b) 1982 – Density 150 kg/m^3 as collected

Figure 6 *Constituents of domestic waste, 1940 and 1982 (after Jarrett, 1979)*

The settlement behaviour of poorly compacted, unsaturated fills tends to be as if they were loose granular materials. All fills settle under their self-weight, the amount depending on the nature and state of compaction of the materials, the age of the fill and the effects of saturation. Imposed loading causes additional settlement.

With modern techniques and by improvement over limited areas and to limited depths, filled land can be rendered suitable for building developments that would have been inconceivable until recently. However, the design of development works must take account of the large vertical movement of deep fill material under its own weight and, to a lesser degree, under the superimposed structural loading.

It is the long-term settlement of the fill rather than its bearing capacity which usually poses the major constraint to its potential for building development. Even with relatively high imposed vertical stresses in deep fills, the principal cause of settlement may not be the structural loading. The long-term movement of a deep fill under a structural load is very largely dictated by the self-weight settlement of the fill mass itself, unless this has already largely taken place. Bearing capacity is not, in fact, the over-riding factor in the engineering of building foundations on fill. Provided settlement is made the design criterion, it is unlikely that bearing capacity will have a determining influence.

11.3.2 Settlement of unsaturated fill under self-weight

Meyerhof (1951) has published curves (Figure 7) showing the long-term settlement with respect to time for various types of unloaded compacted and uncompacted natural inert fills, together with one curve for a well-compacted mixed refuse. Although many times more compressible than even an uncompacted natural fill, the refuse settles in a similar manner. They all exhibit a long-term 'creep' settlement at a rate which decreases exponentially with time elapsed since the deposition of the fill.

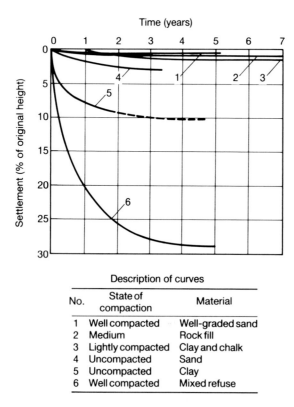

No.	State of compaction	Material
1	Well compacted	Well-graded sand
2	Medium	Rock fill
3	Lightly compacted	Clay and chalk
4	Uncompacted	Sand
5	Uncompacted	Clay
6	Well compacted	Mixed refuse

Figure 7 Observed settlement of fills (after Meyerhof, 1951)

Meyerhof's findings are supported by studies by Sowers (1973), Kilkenny (1968), Charles (1973) and others, which show that the self-weight settlement of an unsaturated fill may be regarded as the result of two consecutive movements:

- Immediate 'primary' compression
- Long-term 'creep' settlement.

Primary compression is found to take place rapidly and to be virtually terminated upon completion of deposition and may usually be disregarded for purely self-weight settlements.

By contrast, creep settlement can continue for many years at a steadily diminishing rate. It is important, and design must take account of the probable creep during the life of a building.

Under conditions of constant effective stress and moisture content, the creep settlement of many fills is approximately linearly related to the logarithm of time elapsed as shown in Figure 8.

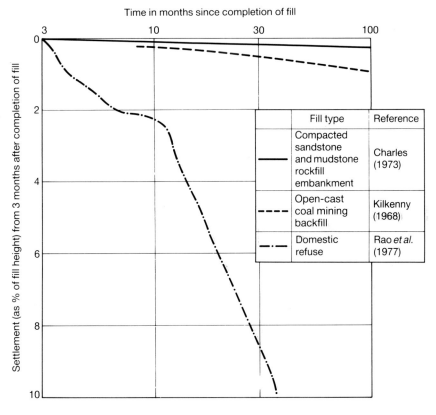

Figure 8 *Creep settlement of fill under self-weight*

11.3.3 Settlement of unsaturated inert fill under structural load

Increased rate of settlement is to be expected when a structural load is imposed on a fill already settling under its own weight. As with self-weight settlement, the concepts of 'primary' compression and long-term 'creep' occurring consecutively also apply.

Primary compression takes place during the construction of the building, and its contribution in this case may be of some importance to the overall settlement suffered by the building.

The long-term creep settlement is generally the more serious problem affecting the structure, especially if the self-weight settlement is still continuing. However, the most significant effect on the structure is likely to be the differential settlements which arise out of the unpredictable and heterogeneous nature of fills. Differential settlement is generally accepted as being up to 75% of the total movement of a building. With large settlements inherent in fills, differential settlement can potentially cause serious distress in buildings.

11.3.4 Collapse settlement

A further potentially serious problem arises from the possibility of a rapid 'collapse' settlement upon inundation of a previously unsaturated, loose fill. This is most severe upon the first inundation, which can occur at any time, irrespective of the time elapsed since the deposition of the fill. Collapse settlement is believed to be due to weakening and crushing of the points of contact between 'granular' components of a fill, and is considered to account for failures of several structures built on apparently stable fills.

The phenomenon can be caused by a rise of the groundwater table into a loose fill, which has previously been unsaturated during its whole life. It is not a common phenomenon, but it should not be assumed that a site is completely stable and safe for development without investigation of the groundwater regime, even though the creep settlement may have diminished to very small values some years after deposition. In addition, a previously unsaturated backfill, which was not systematically compacted during placement, should always be regarded as susceptible to partial or complete collapse upon wetting by percolation from the surface. This risk is particularly possible whenever building developments require deep trenches to be cut through the surface crust.

Measurements by Charles *et al.* (1984) at Horsley opencast coal site are given in Figure 9 and demonstrate a clear relation between rising water table and collapse. The backfilling with sandstone and mudstone fragments of the 70-m deep site was completed by 1970 and site pumping stopped in 1973. With the water table re-established at equilibrium by 1977, the average settlement was slightly less than 1%, measured over the full depth of inundated backfill, with vertical compressions reaching a maximum of 2% locally. In a similar fill of stiff clay lumps, inundated through surface trenches, the corresponding maximum compression locally reached 6% (Charles, 1984).

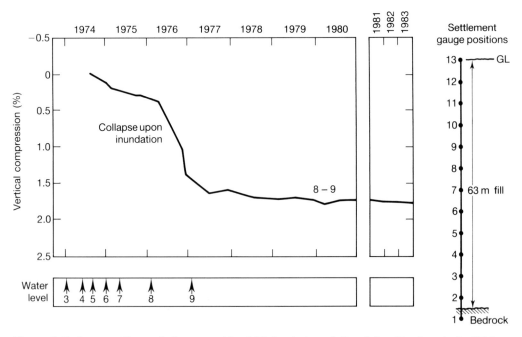

Figure 9 *Collapse settlement of opencast backfill due to inundation (after Charles et al., 1984)*

The effects of inundation on buildings, supported directly on fill, are illustrated in Figures 10(a) and (b) which show schematically the relative settlements suffered by buildings constructed before and after collapse settlement of the fill. In cases where the groundwater regime indicates a risk of future inundation to higher levels than those already reached, the possibility of future significant collapse settlement may counsel against an immediate building development. In such cases it might prove economic to delay construction until it is known that there is little or no potential for collapse settlement to take place. Section 17.3.7 deals with decision making for sites at risk from potential collapse.

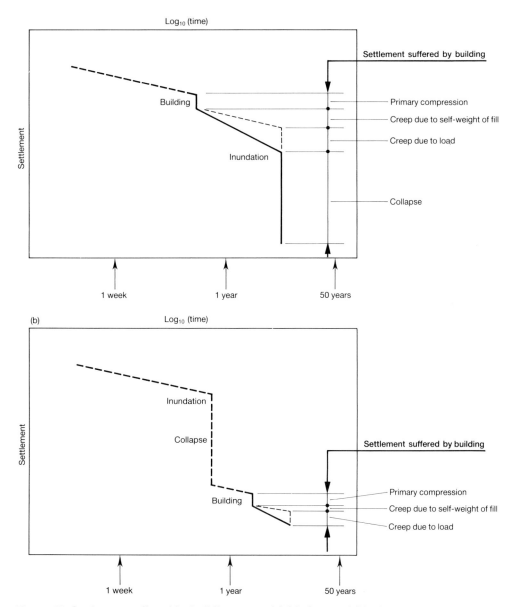

Figure 10 *Settlement suffered by building erected (a) before and (b) after collapse of fill*

11.3.5 Settlement of biodegradable fills

Figure 6 shows how the ash content of domestic refuse has diminished since 1940, with consequent deterioration in the engineering properties of the fill. Settlement of these fills follows the general principles established for inert fills, but is usually more pronounced due to the reactive content of the fill, as well as its normally greater heterogeneity and poorer compaction.

The long-term settlement of refuse is aggravated by the progressive formation of voids due to decay of biodegradable materials and/or underground combustion. Voids may also be created by corrosion or collapse of empty containers or the weakening of local lattices of timber or other materials. The subsequent collapse of such voids increases settlement of biodegradable fills, an effect which is almost impossible to predict.

11.3.6 Consolidation of soft cohesive deposits

In contrast to the loose fills created by tipping, soft cohesive deposits of low permeability result from the deposition of fine materials in water (for example, in tailings lagoons and coal-washing areas). Settlement in such cases (whether due to self-weight or applied load) is controlled by a consolidation process in which excess water pressures dissipate slowly as water is squeezed out of voids of the fill.

Both the magnitude and the rate of settlement can be estimated from consolidation theory by using values of compressibility measured in laboratory tests and field measures of permeability. Geotechnical investigations of a number of such fills have been carried out (Ball, 1979; Krizek and Salem, 1977; Somogyi and Grey, 1977). Thus prediction of settlement is not a problem, but these fills generally have low bearing capacities and may be subject to liquefaction. A firm crust may form over the surface of the lagoon deposit but this may be thin and may overlie very soft material, which can give rise to difficulty of access, particularly for heavy plant.

Very few sites with soft fills have been reclaimed for building development.

11.4 Problems of slags, mining spoils and building wastes

Some types of metallurgical slag, mining spoil and demolition rubble, used as fill material or found deposited in waste areas of old works, can cause problems through chemical instability. Disturbance can lead to weathering and oxidation in the ground, accompanied by volume changes which can cause mechanical disruption of foundations, services and buildings. Such materials or their reaction products, especially sulphates and acids, can further react with concrete, resulting in additional chemical disintegration.

Routine chemical analyses may not always indicate the potential for expansive activity. It is important, therefore, to recognise the metallurgical/mineralogical/chemical nature of the source material and the reactions that it may undergo, and to specify the analyses accordingly.

Sherwood (1987) gives valuable information on most types of waste likely to be encountered as engineering fills. The following notes are intended to supplement Sherwood, regarding the specific phenomena of chemical instability and expansion in the more commonly used spoils. Bridges (1987) describes many slag deposit sites in the UK and abroad and their reclamation, including re-vegetation for ground cover.

Among the more common materials to be treated with caution are:

- Sulphate-bearing building wastes
- Most older iron-making (blast-furnace) slags
- All steel-making slags
- Some colliery spoils and other pyritic shales.

Building wastes. Sulphate-bearing wastes include plaster board, wall plaster, Keene's cement and similar products, generally based on gypsum. Expansion may not be a problem but these materials are well documented as sources of sulphate attack on concrete (BRE, 1975). In acidic ground they can also generate hydrogen sulphide under anaerobic conditions. They are not recommended for use in direct contact with concrete or in situations where troublesome sulphate migration is possible.

Iron- and steel-making slags. Until the decline of iron and steel making in the UK, some seven million tonnes of slag were used annually in civil engineering, the majority of it from blast furnaces. Modern blast-furnace slag, produced during the past forty years and satisfying the provisions of BS1047: Part 2 (1974) *Specification for aircooled blast furnace slag*, is generally an excellent engineering material of fairly constant composition and with total sulphur below 1.5%. Practically all current production is usefully consumed. It is in fact, today, too valuable a material for use as bulk infill.

In contrast, blast-furnace slags of earlier date, as well as practically all steel-making slags, have serious chemical instability defects (Sections 11.4.2 and 11.4.3). They are variable in composition and physical form, and are often of inferior mechanical properties. Upon disturbance, these slags have undesirable weathering and expansion potential, and they should not be accepted under buildings. They are often found as massive slag banks, either buried or exposed, at the sites of old works. Where encountered *in situ*, they should not be disturbed except to remove them completely from the site.

Thomas (1983) and Barry (1985) deal at some length with the problems of reclamation of old iron and steel works.

Pyritic shales. Oxidation of these materials can lead to complex reactions in the ground, and is considered separately in Section 11.4.3.

11.4.1 Blast-furnace slags

With older blast-furnace slags the most serious problems derive from their sulphur content. BS1047: Part 2 (1974) *Specification for aircooled blast furnace slag* sets nominal upper limits of 0.7% sulphate and 2% total sulphur, as a safe margin below which the 'sulphur unsoundness' reaction will not be troublesome. Most modern blast-furnace slags are safely below these values.

With older slags of high sulphur content, weathering can provoke a reaction within the slag mass between the lime-alumina glass, sulphate and water. The resulting sulphoaluminate phase is accompanied by a substantial expansion of about 120% of the original glass volume (Thomas, 1983). The consequent heave may be sufficient to cause mechanical damage to an overlying building. In addition, free sulphate in close contact with concrete can provoke a similar reaction with Portland cement, leading to progressive chemical disintegration of the concrete.

Once a high-sulphur blast-furnace slag is disturbed there is little chance of inhibiting these reactions. Such slags are not normally acceptable as hardcore or fill directly below buildings.

11.4.2 Steel-making slags

Most steel slags are today basic slags containing 'free' lime and very little sulphur. The lime may hydrate with a 100% volume increase (Thomas, 1983). Where the lime is embedded in an otherwise stable slag, impenetrable to moisture, it may remain unreacted for years, until the slag is mechanically disturbed. A similar but more serious expansion may result from the hydration of magnesium oxide (periclase), which can persist longer than lime before hydrating, and can cause disruption many years after the deposition of a fill.

Hydration of free lime or periclase is almost impossible to inhibit. Expansion problems can be expected to be more prevalent than with blast-furnace slags; steel slags are not normally acceptable under buildings.

11.4.3 Oxidation of colliery spoils and other pyritic shales

Basic concepts. The slow oxidation of pyrite-bearing materials (e.g. some colliery spoils) can give rise to troublesome ground acidity, as well as sulphate formation with associated volume changes. Where these materials underlie foundations, the ensuing damage is found almost invariably to have been caused by chemical attack of sulphate on concrete. The presence of secondary sulphate products within the cracks of failed concrete slabs may be evidence of such chemical attack (BRE, 1977, 1981, 1983b, c).

Some naturally occurring pyritic shale formations have been known occasionally to cause mechanical damage to buildings founded directly on the rock, through volume changes on oxidation of the pyrite (Hawkins and Pinches, 1987). Nixon (1978) reported similar damage through expansion of fill materials 'borrowed' from one such shale formation. He also included a provisional list of naturally occurring shale formations to be treated with suspicion on account of their expansive potential. A more extensive list was later produced by Taylor and Cripps (1984).

This type of mechanical damage, due to expansion within the pyrite material, appears to be an uncommon phenomenon. Nixon (1978) stated that there was no documented evidence of such damage being caused by colliery spoil used as fill in the UK, whereas Taylor and Cripps (1984) included some of the Coal Measures in the cautionary list. Other recent studies (Wilson, 1987; Taylor, 1988) also appear to suggest that spoils from certain of the Coal Measures may be suspect. Pending more definitive documentation, it would be prudent to accept the possibility, albeit remote, of expansive damage also from colliery spoil.

The increased ground acidity associated with pyritic oxidation is harmful to plant life, and tends to enhance metal mobility with a resulting increase in metal contamination in drainage water.

Oxidation cycle. The progress of oxidation through a pyrite-bearing mass in the ground is illustrated in Figure 11. This figure, and the explanatory notes which follow, are intended as a simplification of a complex phenomenon, and may not correspond in chronological detail with any particular case history. In broad terms, however, Pulford and Duncan (1978) have suggested that it might be possible to assess the harmful acid potential of a given coal waste by identification of the stage reached in the oxidation cycle (by measuring pH and acetic-acid-soluble sulphate levels).

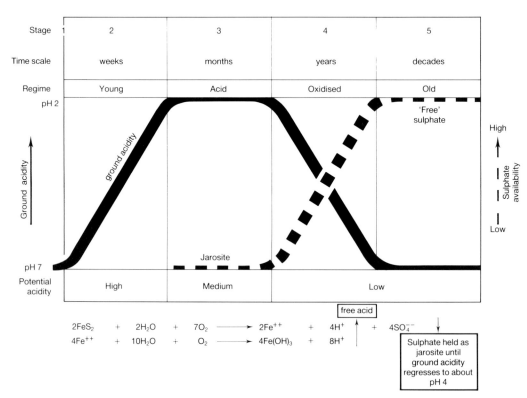

Figure 11 *Acidity and sulphate production from oxidation of pyrite*

Acid production. Freshly excavated colliery spoil, containing completely unoxidised pyrite, may have a neutral or slightly alkaline reaction (Stage 1). Oxidation quickly ensues upon exposure to the air in the presence of water and bacteria (especially in acid conditions), with the formation of sulphate and free acid (Stage 2). Vegetation soon suffers from the increasing acidity, but attempts at artificial neutralisation in the early stages are usually abortive, due to the high acidity potential remaining in the ground from still-unoxidised pyrites. As the oxidation progresses, the acidity of an increasingly large ground mass reaches about pH2. Thus the acid tends to peak (Stage 3) and then slowly to leach away (Stage 4), and the ground returns towards neutral after a period of years or decades (Stage 5).

Sulphate production. In the early stages, once the pH drops below about 4, the sulphate is held as the insoluble mineral jarosite, $KFe_3(SO_4)_2(OH)_6$. The formation of jarosite is accompanied by an expansion of 115% of the original volume of crystalline pyrite (Taylor and Cripps, 1984).

The return of pH towards neutrality during Stage 4 coincides with the liberation of the sulphate held as jarosite, and its reactivity with other available minerals. Sulphate availability thus tends to peak later than acidity. If calcite is present gypsum is likely to form, with a volume expansion of 103% over the calcite volume (Taylor and Cripps, 1984).

In the few cases where expansion has led to building damage on fills or formations of pyritic materials, gypsum production is usually regarded as the critical factor. Gypsum has been found crystallised in fissures in expanded bedrock (Hawkins and Pinches, 1987) and in laminations in the shale components of expanded fill (Nixon, 1978). It may also migrate in acid solution and crystallise elsewhere.

12. Hazards of contamination

This section applies to both industrially contaminated and fill sites. Problems arise out of the chemically and biochemically reactive natures of the dangerous materials, residues, by-products, etc. present in derelict sites which have been contaminated, either by industrial or other activities carried out on the site or by harmful materials deposited as fill.

Derelict land can contain massive obstructions to boring, pile-driving and earth-moving operations. For convenience, these are also dealt with in this section.

12.1 Type and content of contaminated sites

To a varying degree, the sites of past industries are contaminated with residues, wastes and by-products. They may also be congested with old buildings, foundations, etc., and machinery and industrial ancillaries, which may themselves be sources of contamination.

The land may be contaminated by atmospheric fall-out; by leakage, flooding, spraying or seepage of liquids; by deposition or burying of solids as in waste disposal; or by deposition of sludges, either on the surface or in pits or lagoons. All former industrial and wastefill land should be regarded as potentially contaminated.

Table 4 lists some of the more common types of site likely to be contaminated. In addition, even since the Control of Pollution Act 1974, there may be danger of localised contamination in an apparently greenfield site as a result of illegal fly-tipping.

Table 4 Examples of contaminated sites

Landfill and other waste-disposal sites
Gasworks, coal-carbonisation plants and ancillary by-product works
Sewage works and farms
Scrapyards
Railway land, especially large sidings, depots and breaking yards
Roads, airports and abandoned wartime airfields
Docks, canals and abandoned or infilled port ancillaries; ship-breaking yards
Oil refineries, petroleum storage and distribution sites
Metal mines, smelters, foundries, steel works, metal-finishing works
Mineral extraction sites not yet infilled (quarries, coal mines, clay pits, tin and china clay mines, etc.)
Glass works
Chemical works
Munitions production and testing sites, wartime installations
Asbestos works, and buildings incorporating asbestos
Tanneries and fellmongeries
Paper and printing works
Industries making or using wood preservatives, herbicides and pesticides
Cotton and other textile mills and bleach works
Metal-plating works and yards
Paint-manufacturing works
Brickworks, potteries and ceramic works
Nuclear power stations, radioactive storage/disposal installations

12.2 Problems from contaminated site hazards

Of the hazards listed in Table 1 (page 21), those most likely to be encountered on contaminated sites comprise:

Physical
- Obstructions

Chemical/biochemical
- Health and environmental hazards
- Combustibility
- Aggressive attack
- Gas emission
- Leachate production.

Physico-chemical
- Expansive reactions.

On contaminated sites the problems of physical obstructions arise from buried remains of old industrial or commercial developments. Similar problems may occur on fill sites from massive blocks of concrete, rock, building rubble and slag deposits. Obstructions are discussed briefly in Section 12.3.

Generally, it is the chemical contamination hazard which gives rise to the most serious problems on old industrial and commercial sites. Some significant contaminants which may be encountered are given in Table 5. The list is not exhaustive, and the contaminants actually present will depend on past use(s) of a site. On the other hand, not all the contaminants in Table 5 will be of significance for a particular end use. (Table 15 on page 92 gives a restricted number of 'listed' chemical hazards most likely to require attention on the majority of sites.)

Table 5 Some significant contaminants

Metals and their compounds: arsenic, barium (soluble), beryllium, boron,* cadmium, chromium, copper,* iron,* lead, manganese,* mercury, molybdenum,* nickel, selenium, silver, thallium, zinc*

Non-metals: chlorides, sulphides, sulphates, sulphur

Acids: hydrochloric, phosphoric, sulphuric

Alkalis: caustic solutions, ammoniacal liquors

Organic substances: phenols, cyanides (free and complex), thiocyanates, hydrocarbons, oils, tarry wastes, polychlorinated biphenyls (PCBs), pesticides, herbicides and other chlorinated hydrocarbons, polynuclear aromatic hydrocarbons (PAHs)

Putrescible, biodegradable matter: domestic waste, food and vegetable residues, paper, packaging

Miscellaneous materials: asbestos, radioactive substances, glass, rubble, coal wastes, pyritic shales, methane

---

* In trace amounts, essential to plant and animal health.

Aggressive action and corrosion may arise when contaminants, principally in solution, come into contact with building materials. Sulphate attack on concrete, for example, is fairly well researched though not all factors are fully understood. Effects of other contaminants on building materials are not so well known and documented.

Hazards to human and animal health and well-being may arise directly from inhalation, ingestion or contact with toxic or harmful contaminants. Additionally, take-up of a contaminant by a plant may be harmful to the plant, or indirectly to humans and animals either feeding on the plant itself or on other animal consumers of the plant in the food supply chain.

Of the metals in Table 5, those marked with an asterisk are essential in trace quantities for the health of plants and animals. For these there is very little margin between beneficial and harmful concentrations.

Metals in the ground do not normally present a hazard for site workers (except in the form of dust), but subsequent occupants or site users may be at greater risk through more prolonged exposure. On the other hand, oils and tars will probably be a greater hazard to site workers than to later occupants who are less likely to come into contact with them.

Metallic contaminants can be a problem to metallic building materials through electrolytic action. Dissolved salts can similarly cause problems.

12.3 Obstructions

The majority of contaminated sites have, by virtue of past usage, accommodated buildings, ancillary structures and underground services. These present potential obstructions to redevelopment which, if not foreseen and planned for, can have major significance when discovered during construction. Figure 12 illustrates an actual record from a gasworks site.

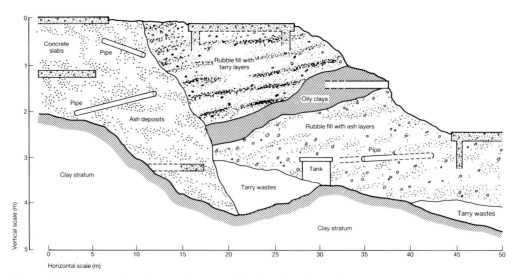

Figure 12 *Cross-section of part of a gasworks site, Gateshead (after Cairney, 1985)*

It is an unfortunate fact that the records of old buildings are quite often unavailable or have insufficient detail for location and sizing of the underground features. The technical problems posed by the underground remains of old structures include:

- Obstructions to piling
- Obstructions to ground treatment (stone columns, grouting)
- Concentration of contaminants (e.g. tar wells, pits)
- Dispersion of contaminants (leaking pipes)
- Hard spots under foundations and cover layers
- Obstructions to service runs
- Interference with site investigation.

Additional problems arise if the buildings have been demolished in that visual evidence of underground obstructions has usually been obliterated. Where old basements have been infilled carelessly, the sudden change of stiffness from the walls to the loose and often mixed fill in the old void gives rise to problems in the design of new foundations.

Likewise, fill sites may also contain serious obstructions to piling, ground treatment and site investigation in the form of massive concrete waste, building and demolition rubble, timber, drums and containers, quarry waste, broken rock (on opencast mining sites), etc. Deposits of slag on iron and steelworks and some fill sites may also hold obstructions from large individual pieces of slag (Figures 13 and 14).

12.4 Health and environmental hazards: general

12.4.1 Introduction

Common contaminants, listed in Table 5, represent hazards to the health and safety of human, animal and plant life. In humans, the effects may be short or long term, and vary from simple irritation to extreme toxicity. In all cases the sensitivity of the target to the particular hazard is of crucial importance.

The hazards posed by harmful contaminants depend on five main factors:

1. Concentration of contaminant
2. Availability of contaminant to target
3. Exposure time of target to contaminant
4. Route to target
5. Target sensitivity.

It is self-evident that the hazard represented by a contaminant increases as concentration increases but the other four factors require more explanation.

Figure 13 *Massive foundations and cast iron gas pipes (photograph by courtesy of Environmental Safety Centre, Harwell Laboratory)*

Figure 14 *Underground gas holder (photograph by courtesy of Environmental Safety Centre, Harwell Laboratory)*

12.4.2 Availability of contaminant

The toxicity of a contaminant depends on its chemical form. Thus, free cyanide (as in soluble cyanides such as potassium cyanide, KCN) is much more toxic than a complex cyanide such as potassium ferrocyanide, $K_4Fe(CN)_6$. The environment in which a contaminant resides also affects its potential as a hazard. Thus metals in acid soils tend to be more available to plants than those in alkaline soils.

Persistence in the ground also affects availability. Some contaminants (for example, phenols and, to a lesser extent, free cyanide) are biodegradable and their concentrations tend to diminish with time. Likewise, soluble contaminants may be leached out by migrating groundwater. In putrescible fills progressive biodegradation also leads to loss of organics into the atmosphere, and precipitation of heavy metals as carbonates, hydroxides and sulphides (Section 12.15). These factors may lead to a relative increase in the total heavy metal content, with diminution in the available (soluble) metal concentration.

Changes in the water table or temperature or exposure of contaminants to the atmosphere during site operations may result in physical or chemical reactions which may change the concentration or availability of contaminants.

12.4.3 Exposure time

The effect of a chemical absorbed and retained within the body may be acute, resulting from short-term exposure, or chronic from long-term exposure to lower concentrations. Acute exposure may also show delayed effect (e.g. 20–30 years in the case of asbestos).

Acute poisoning hazards (from materials such as free cyanides, arsenic, phenols and, to lesser degree, sulphates) are of prime concern to the safety of site workers who may risk exposure for short periods to relatively high concentrations. Safety precautions and site working procedures require careful planning (Section 23).

Chronic hazards (from such contaminants as arsenic, phenols, some hydrocarbons and polychlorinated biphenyls (PCBs) and heavy metals) are more likely to affect later residents and long-term occupants of the land. However, assessment of the long-term threat from toxic materials is difficult due to the variability in the long-term reaction of individuals, the very low concentrations and long time periods involved, and to incomplete knowledge of the pathological effects of long-term exposure.

12.4.4 Route to target

The possible routes by which a contaminant may reach a target comprise:

- Inhalation
- Ingestion
- Contact.

Contaminants dangerous by one route may be relatively innocuous by another. Thus asbestos, which is a known carcinogenic hazard when inhaled as a dust, is less dangerous when ingested.

12.4.5 Target sensitivity

Targets likely to be affected by contamination in the ground may be at risk either because of their greater than normal chance of exposure or because of their special sensitivity. Children are at risk on both scores.

Table 6 illustrates the relative vulnerability of targets in terms of exposure risk and sensitivity to contaminants.

12.5 Health hazards: inhalation

Hazards through inhalation involve either gases or particulate materials in the form of dusts (greater than 10 micron size) or fumes (less than 10 micron size).

Table 6 Sensitivity of targets on contaminated sites

Target	Exposure risk	Relative vulnerability
Humans[1]		
Children[2]	High[2]	High[3]
Adults	Low	Low
Gardeners (full time)	High	Low
Elderly and infirm	Low	High
Own-food growers[5]	High	Low
Site workers	High	Low
Animals[1]		
Herbivores	High[4]	High
Fish	High	High
Plants[5]		
See Sections 12.6.2 and 12.8		

Notes:
1. Humans and animals mainly at risk through ingestion or contact.
2. Children are exposed through careless dietary habits and mode of life.
3. Children are more sensitive than adults.
4. Grazing animals ingest appreciable quantities of soil.
5. Many edible plants are able to take up and accumulate metals (without apparent harm to the plant) in concentrations which would be dangerous to humans and animals eating the plant if ingested over a sufficiently long time as a major component of the diet. On the other hand, many edible plants act as 'geochemical barriers' which protect human or animal consumers from the full effect of the metal concentration in the ground (Thornton, 1985).

12.5.1 Dusts and fumes

These normally create a short-term hazard during earth-moving operations, mainly affecting site workers and occupiers of nearby property. Long-term hazards may result from persistent toxicants (such as dust from lead and other metals, coal tars and asbestos) blowing from a contaminated site even for only a limited period, or from less toxic dusts over a long period. The quantity of dust in the air will increase with decreasing moisture content and particle size. Water-spraying of dusty areas will normally be required during site works. It may also be necessary long-term if the site is to be abandoned (Section 7.4).

The particle size of the contaminant may significantly affect the route and magnitude of uptake by the body. Fumes can penetrate deep into the lungs, whereas many dusts will be trapped in the nose and throat. Some dusts, however, do penetrate and may be particularly hazardous. For example, pitch (found on tar and by-product works) and asbestos are known respiratory carcinogens.

12.5.2 Gases

Gases may be produced on derelict sites by decay, vaporisation, chemical reaction or combustion:

Aerobic and anaerobic decay—produces landfill gas (mainly methane and carbon dioxide)

Vaporisation—principally of spilt fuel, dumped solvents or hydrocarbon wastes.

Chemical reaction
- Production of hydrogen sulphide from sulphur compounds (for example, sulphides, sulphates and organics, gypsum and plasterboard)
- Various reactions arising from co-disposal (for example, of domestic refuse and solvent)
- Acid/alkaline leachate reacting with deposited chemicals (for example, cyanide-producing hydrogen cyanide)
- Limestone/chalk reacting with acid to produce carbon dioxide.

Combustion—for example, sulphur dioxide from spent oxide, thiosulphates, and sulphurs; hydrogen cyanide from thiocyanate and free cyanides; carbon dioxide from burning organic materials; carbon monoxide from smouldering underground fire with restricted oxygen supply; dioxin and other carcinogens from burning plastics, transformer oils, etc.

Hazardous gases may be present on a derelict site as a result of continuing production, or may have accumulated over years in pockets which, once pierced, will become exhausted.

Gas production may also be triggered by disturbance of the site. Excavations of considerable extent have been known to fill with gas following a reduction in barometric pressure. Gases may also dissolve in groundwater and come out of solution on reduction of atmospheric or overburden pressure. The characteristics and effects of some hazardous gases likely to be present on a derelict site are given in Table 7.

Table 7 Characteristics and effects of hazardous gases

Gas	Characteristics	Effect	Special features
Methane	Colourless Odourless Lighter than air	Asphyxiant	Flammable limits 5–15% in air Can explode in confined spaces Toxic to vegetation due to deoxygenation of root zone
Carbon dioxide	Colourless Odourless Denser than air	Toxic asphyxiant	Can build up in pits and excavations Corrosive in solution to metals and concrete Comparatively ready soluble
Landfill gas	Colourless	Toxic asphyxiant	Mixture mainly of methane and carbon dioxide Properties as main constituents but not odourless Not always lighter than air
Hydrogen sulphide	Colourless 'Rotten egg' smell Denser than air	Highly toxic	Flammable Explosive limits 4.3–45.5% in air Causes olfactory fatigue (loss of smell) at 20 ppm Toxic limits reached without odour warning Soluble in water and solvents Toxic to plants
Hydrogen	Colourless Odourless Lighter than air	Non-toxic asphyxiant	Highly flammable Explosive limits 4–74% in air
Carbon monoxide	Colourless Odourless	Highly toxic	Flammable limits 12–75% in air Product of incomplete combustion
Sulphur dioxide	Colourless Pungent smell	Respiratory irritation Toxic	Corrosive in solution
Hydrogen cyanide	Colourless Faint 'almond' smell	Highly toxic	Flammable Explosive
Fuel gases	Colourless 'Petrol' smell	Non-toxic but narcotic	Flammable/explosive May cause anoxaemia at concentrations above 30% in air
Organic vapours (e.g. benzene)	Colourless 'Paint' smell	Carcinogenic Toxic Narcotic	Flammable/explosive Can cause dizziness after short exposure Have high vapour pressure

Note: all gases are at least partially soluble in water and come out of solution on reduction of atmospheric pressure.

Some gases, although relatively harmless, have offensive odours. Exposure to malodour may prove unacceptable in the long term for people living and working near to the source, or in the short term for workmen on the site.

The particular case of landfill gas is discussed in detail in Section 12.11. The hazards of landfill gas are generally the same as for its principal constituents, methane and carbon dioxide.

12.6 Health hazards: ingestion

12.6.1 Hazards and targets

There are three main routes for the ingestion of contaminants present on a derelict site:

1. Direct ingestion of contaminated soil
2. Consumption of plants (or animals) which have taken up or ingested toxic contaminants
3. Ingestion of contaminated drinking water.

As regards direct ingestion of contaminated soils the most sensitive targets are children, but all 'pica' subjects (habitual eaters of soil and other non-food materials) are at risk. Grazing animals are similarly at risk.

In the case of uptake of toxic materials by edible plants, the most sensitive targets are allotment holders and others whose diet includes a considerable amount of home-grown food. Commercially grown crops should not normally be a major hazard given the dilution effect of normal wholesale distribution.

Contamination of water includes the pollution of:

- Surface water (ponds, streams and standing water)
- Water piped across contaminated ground
- Groundwater or aquifers.

Contaminated water is of greatest danger to children, animals and fish. In a lake or pond, it affects an amenity. Water contaminated via pipes could put many people at risk. More serious would be aquifer contamination, affecting the whole process of water supply, treatment and use. Even at contaminant concentrations well below toxic levels, the aesthetic quality of the water (taste, smell, etc.) may not be acceptable.

12.6.2 Uptake of contaminants by edible plants

A risk to health may be posed by certain edible crops which are able to absorb sufficient metal to be a hazard to humans and animals without apparent harm to the plants themselves. The metals of most concern are arsenic, cadmium, lead and mercury. Cadmium is known to be fairly readily absorbed by the body and only slowly released, so that it is liable to become a chronic health hazard. Lead and mercury similarly have chronic effects but are believed to be somewhat more readily released from the body.

The uptake and retention of toxic metals by plants depends on two essential factors:

1. *Species and variety*: lettuce is the most efficient accumulator of cadmium likely to affect growers in the UK. Brassicas and radishes, for example are less seriously affected. However, the contribution of a particular plant to a diet must also be considered. Potatoes, for instance, may be less contaminated than lettuce but contribute far more to the annual diet.
2. *Availability of contaminant to plant*: this is dealt with in more detail in Section 12.8.

Recent work by Lyons (1983) and Thornton (1985), however, suggests that some plants tend to act as geochemical barriers which protect humans from the full effect of metals in the soil. A proportion of the metal found in or on vegetation may also be from airborne surface contamination. Thus the lead level in and on lettuces has been found to decrease exponentially with distance from the centres of large towns.

12.6.3 Contamination of water resources

Contamination on a derelict site may pose a threat to piped drinking-water supplies and to ground and surface water resources. Drinking water piped through contaminated ground may be at risk from defects in the pipes or joints arising from aggressive attack or mechanical failure. Moreover, certain organic compounds (phenols and solvents, for example) can diffuse through plastic pipes or joints sufficiently to taint the water without damage to the pipe.

Wells drawing from local contaminated perched water tables may be severely affected, even though the main aquifer may be unharmed. Gas works pollution, for example, has on occasion enforced the closure of pumped drinking-water wells.

Site works may aggravate water-contamination problems. Stripping of surface soils may lead to gross pollution of surface waters. Excavation of contamination below the water table will inevitably create large volumes of contaminated water which may require extensive treatment before disposal. Similarly, the breaching of buried tanks or pipes could release contamination into the local groundwater.

Some of the more common contaminants which may affect groundwater are shown in Table 8.

A special problem arises from the relatively non-toxic thiocyanates which, in contact with chlorinated drinking water, can produce toxic cyanogen chloride.

Table 8 Contaminants which may affect groundwater

Solubility	Contaminant	Notes
Soluble	Free ammonia	1
	Sulphates	
	Chlorides	
	Phenols	1, 2, 3
	Free cyanides	
	Complex cyanides	1
	Polychlorinated biphenyls (PCBs)	4
	Thiocyanates	5
Low solubility	Tar acids	
	Coal tar derivatives	
	Solvents	
	Hydrocarbons	
	Chlorinated hydrocarbons	
	Polynuclear aromatic hydrocarbons (PAHs)	
	Pesticides, herbicides	
	Metals	
	Arsenic	

Notes: 1. Though non-toxic at very low concentrations, can affect the taste of water to an unacceptable degree.
2. Can affect the taste of chlorinated water by reacting to produce chlorphenols and chlorcresols, which are non-toxic.
3. Biodegradable: diminishing hazard.
4. Persistent in environment: enduring hazard to food chain through aquatic life.
5. Relatively non-toxic but react with chlorinated water to produce toxic cyanogen chloride.

12.7 Health hazards: contact with contaminants

12.7.1 Hazards and targets

Direct skin contact with contaminants may lead to:

- Chronic skin effects (dermatitis)
- Acute skin irritation
- Harmful effects due to absorption.

Sensitive targets for skin contact are:

- Site workers
- Children
- Gardeners.

Site workers are likely to be exposed for the relatively short duration of the site investigation and reclamation works, but the hazards may be intense. On the other hand, unless the contamination is remedied during the site works a chronic hazard is likely to remain for those most frequently handling soil (children, gardeners, and especially maintenance workers) after completion of the development. Children may also be at risk prior to development if the site is accessible to them.

12.7.2 Skin contact

Site workers may come into contact with residues and other substances remaining on the site from previous usages. Liquid residues may be found in tanks and pipes, in sumps or underground cavities, or as contaminated surface water in trenches and excavations. Residual liquids or sludges may be present as a complex mixture of contaminants.

Construction workers are exposed to an increasing range of chemicals used, for example, in glues, sealants and surface coatings. They may also already be suffering from chronic skin complaints (e.g. cement dermatitis—from the alkalinity and chromium content of cement) which may increase their sensitivity to exposure to other chemicals.

Environmental Resources Ltd (DoE, 1987) detail the potential toxic hazards of skin contact with the principal liquids to be encountered on former coal carbonisation works, i.e. ammoniacal liquors, phenols and pumpable coal tars. Skin contact is also undesirable with acid and alkaline solids or liquids, such as acid deposits of spent oxide on gas works sites, and alkaline slags from iron and steel making.

Some carcinogens, such as coal tars and phenols, can also cause less serious short-term symptoms (acute skin irritation and sensitisation). Carcinogenic effects are thought to result only from regular contact over prolonged periods. Other potential carcinogens such as polychlorinated biphenyls (PCBs) are generally without short-term irritant effect.

A number of contaminants may give rise to short-term mild skin irritation, examples being sulphides, sulphates and cyanides. Others, such as lime, cements and diesel oil, can cause chronic dermatitis.

12.8 Phytotoxicity

12.8.1 General

Phytotoxicity (the toxic effect of harmful substances on plant life) is not directly a technical problem affecting building works. It is however, a hazard of importance to sensitive targets ancillary to building works (e.g. food-growing areas, gardens, open spaces, parks, play areas, etc.) where humans and animals may be affected by poisoned vegetation.

The effect of a given chemical contaminant on a plant depends on:

- Sensitivity of the plant to the contaminants
- Ability of the plant to regulate uptake
- Availability of the contaminant to the plant.

Phytotoxic contaminants may prove a hazard to any vegetation, whether for agriculture, horticulture or landscaping around office buildings or industrial units. Death of vegetation is unsightly and expensive and may lead to surface erosion, exposing buried toxic materials. The natural annual cycle of growth and die-back can lead to accumulation of contaminant in the surface litter layer.

12.8.2 Sensitivity

Sensitivity varies with the species, variety and stage of growth of the plant. Metal-tolerant strains (for instance) may evolve, so that the presence of healthy vegetation may not necessarily imply freedom from contamination.

Even in the absence of harmful contaminants, excess or deficiency of essential trace element nutrients can be equally harmful to a plant. Abnormally acidic conditions (say, arising from spent oxide in soil) or alkaline conditions (as from blast-furnace slag in soil) may also damage vegetation. Anaerobic conditions, as in a waterlogged soil, are also harmful.

12.8.3 Availability

A potentially phytotoxic contaminant will have a harmful effect when:

- There is physical contact with plants or trees
- The toxic elements of contaminant are chemically 'available' to the plants.

Physical contact may result from root penetration, groundwater excursions and lateral migration, upward soil suction effects in fine-grained soils, decay of contaminated dead vegetation and activities of earthworms. In fine-grained soils, complex cyanides have been known to rise 2 m under the effect of soil suction.

The combination of chemical and soil influences the availability of a contaminant for uptake by a plant. Metals are more available in an acid soil; alkaline soils tend, instead, to 'fix' the contaminant. Metal availability is also affected by the organic content of the soil, being generally lower when the organic content is high. Availability is also generally lower in clayey soils than in sandy ones.

The common phytotoxic contaminants are boron, copper, nickel, zinc and sulphates. Others include sulphides (where ground conditions favour the formation of hydrogen sulphide) and complex cyanides (as in spent oxide waste). Methane and carbon dioxide (from landfill gas) can also lead to the death of vegetation by de-oxygenating the soil.

12.9 Combustion

12.9.1 General behaviour

If the temperature of a combustible material underground is locally raised to ignition point in the presence of oxygen, an exothermic reaction may start with the release of heat. Smouldering can be started by sustained heat from a source which need not be at a very high temperature. If the rate of heat generation sufficiently exceeds the heat loss, a self-sustaining reaction can continue as an underground combustion, provided air remains available. The air supply is usually so restricted that flaming is unlikely. The reaction therefore proceeds as a slow smouldering, which may continue for years with underground temperatures as low as 400°C, and with little evidence on the surface except for occasional wisps of steam or smoke.

Since heat losses diminish with increasing depth of cover, active fires may be deeply located at 4 m or more below ground level. The progress of the underground smouldering is usually slow, but extremely difficult to locate and extinguish once started. The burning of combustible material leaves underground voids, which may collapse and cause settlement.

Much useful information on underground fire hazards is given in ICRCL Note 61/84 (1986).

12.9.2 Sources of ignition

The heat required to ignite the underground combustible material may have various origins:

- Materials already burning when deposited underground
- Fires at ground level, deliberate or accidental
- Localised heating below ground (for example, by electric cables, boiler houses or furnaces)
- Self-heating within the combustible material itself.

Self-heating is a less common cause of ignition, and is still not a well-documented phenomenon.

12.9.3 Self-heating leading to ignition

Slow oxidation of combustible materials underground, or the biodegradation of organic components of fill, are both exothermic reactions tending to cause a temperature rise within the mass of the material. The temperature reached within the reactant mass depends on the balance between the rates of heat output and heat loss. Other things being equal, the reaction rate (that is, the heat-generation rate) and the heat loss rate both increase with temperature: the reaction rate exponentially and the heat loss rate linearly.

The commonly occurring conditions for a small mass near the surface may be represented as shown in Figure 15(a). At temperature T_a, rate of heat gain exceeds rate of heat loss so that the temperature rises until at point X heat input and loss are equal. T_x represents the modest equilibrium temperature reached by the material.

If, in contrast, conditions were such that heat gain always exceeded heat loss, the temperature would tend to rise continually (as Figure 15(b)); this 'thermal runaway' could continue until ignition. Such conditions might obtain with material buried deeper underground with consequent reduced heat loss. Alternatively, an increase in the size of reactant mass would also tend to favour heat gain over heat loss, since the reaction rate varies as the volume of material, whereas the heat loss varies only as the surface area.

If the self-heating were due to oxidation the temperature could rise up to ignition point, when smouldering would start. Biodegradation, on the other hand, would stop at a temperature of about 75°C, above which the bacteria cannot survive. Self-heating initiated by biodegradation alone is unlikely to cause self-ignition, but can give the appearance of a fire underground, with emission of steam and warm surface temperatures.

Table 9 lists some of the more common materials prone to self-heating under favourable conditions. The tendency to self-heat is most marked in colliery wastes, gasworks spent

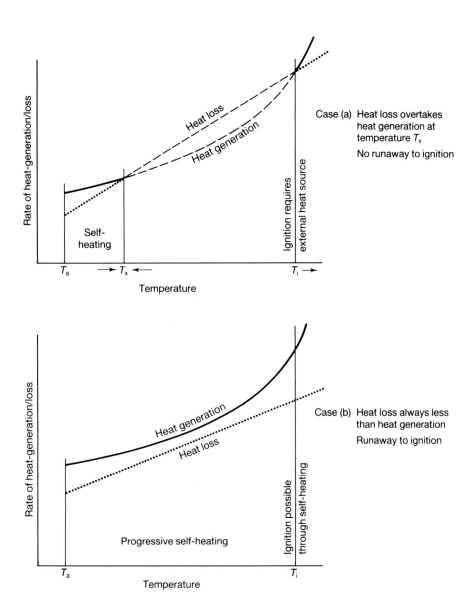

Figure 15 *Effect of self-heating on combustibles*

Table 9 Some self-heating materials (after Beever, 1982)

Animal feedstuffs	Fertilisers	Laundry	Seeds
Beans	Fishmeal	Maize	Seedcake
Bone meal	Flax	Manure	Silage
Brewing grains, spent	Foam and plastic	Milk products	Sisal
Carbon	Free sulphur	Monomers for	Soap powder
Celluloid	Grains	polymerisation	Soybeans
Coal and coal wastes	Grass	Oils and oily wastes	Spent oxides
Copper ore concentrates	Hay	Palm kernels	Straw
Copra	Hemp	Peat	Tarry materials
Cotton	Hides	Plastics, powdered	Varnished fabric
Cottonseed	Iron filings/wool	Rags*	Wood fibreboard
Distillers dried grains	Iron pyrites	Rapeseed	Wood flour
Dried sewage sludge	Jute	Rice bran	Wool*
Fats*	Lagging	Sawdust	Zinc powder

* Some oils and fats are readily oxidised and become hazardous if dispersed on a porous medium such as cloth or powder.

oxide, and exothermic materials from leaking containers. For materials which oxidise, the finer the particle size, the greater the specific surface and oxidation rate, which, in turn, leads to a higher rate of heat generation. Small amounts of water increase oxidation and the heating rate, whereas larger amounts increase the thermal conductivity and heat loss, offsetting any heating gain.

12.10 Fire and explosion

12.10.1 General phenomena

Readily vaporised flammable materials on derelict sites can be the cause of fire or explosion. The properties of twenty such substances which may be found on derelict sites are set out in Table 10.

Table 10 Properties of some common flammable liquids and gases

	Explosive limits (% by volume in air)	Flash point (°C)	Vapour density (air=1)	Maximum explosion pressure (bar)	Ignition temperature (°C)
Acetone	2.6–12.8	−18	2.00	5.7	538
Acetylene (gas)	2.5–81.0	−	0.91	10.3	335
Ammonia (gas)	15.0–28.0	−	0.58	−	651
Benzene	1.4–7.1	−11	2.77	6.7	538
Butane (commercial)	1.8–9.0	−	1.9–2.1	6.7	410
Carbon disulphide	1.25–44.0	−30	2.64	−	110
Carbon monoxide	12.5–74.0	−	0.97	−	510
Ethyl acetate	2.2–11.4	−4	3.04	−	484
Ethyl alcohol	3.3–19.0	14	1.59	6.8	392
Ethylene	3.1–32.0	−	0.98	8.2	450
Ethyl ether	1.85–48.0	−40	2.56	7.2	180
Ethyl nitrate	4.0–100.0	10	3.14	−	85
Hexane	1.2–7.5	−22	2.97	6.3	234
Hydrogen	4.0–75.0	−	0.07	7.0	585
Hydrogen sulphide	4.3–45.5	−	−	−	−
Methane	5.0–15.0	−	0.55	−	538
Pentane	1.5–7.8	−50	2.48	−	287
Petroleum spirit	1.3–6.0	−43	3.0–4.0	−	250–400
Propane (commercial)	2.2–10.0	−	1.4–1.6	6.6	460
Toluene	1.3–7.0	4	3.14	6.3	520
Town gas	5.3–32.0	−	0.46	−	600

A mixture of flammable gas and air may ignite if part of it is raised locally to the ignition temperature. Rapid propagation of flame is accompanied by an even faster expansion of heated gases, creating a pressure wave which travels outwards very rapidly. If these waves are impeded, pressure can build up to over 5 bar (500 kN/m^2) in milliseconds, resulting in the phenomenon of explosion.

12.10.2 Factors affecting explosion risk

Five factors given in Table 10 influence the explosion risks of flammable liquids and gases.

Explosive limits. The speed of burning varies with the relative amounts of fuel and air. If the fastest-burning fuel:air mixture is diluted by either fuel or air a slower burning mixture results, until concentrations (known as the upper and lower explosive limits) are reached at which the mixture will only just burn. For most gases and vapours the lowest explosive limit requires less than 6% of fuel. Readily flammable mixtures can thus form even when only small quantities of fuel are released into the air.

Flash point. The flash point is the lowest temperature at which a flammable liquid gives off ignitable vapour. Liquids with flash points below ambient air temperature can be potential explosive hazards.

Vapour density. Lighter-than-air gases are more readily and safely dispersed than heavy gases and vapours which tend to accumulate and spread at floor level in explosive concentrations. The contents of LPG cylinders (vapour density = 2.11) are particularly hazardous.

Maximum explosion pressures. The pressure developed in an explosion in an enclosed space depends on:

- The fuel concentration. At some point between the two explosive limits the flame-propagation rate is a maximum, and the explosion is the most violent
- The proportion of the space filled by explosive mixture
- The ambient temperature and pressure
- The amount of venting.

Most structural walls are damaged by pressures exceeding 0.2 bar (20 kN/m²).

Ignition temperatures. Temperatures of moderately hot bodies may be above the ignition temperatures of commonly flammable substances, thus:

- Dull red   ~700°C
- Faint red   ~500°C
- Black heat   up to 400°C.

All common industrial gases, including methane, ignite below 700°C and are thus at risk from objects at dull red heat. Gases with ignition temperatures below 500°C, including butane, can be ignited by surfaces at black heat.

12.11 Gas emission

12.11.1 Background

The hazards of inhalation, flammability and explosive potential of gases in general on derelict sites have been discussed previously. Particular problems arise from the anaerobic biodegradation of organic materials, which produces a mixture of gases known as landfill gas. It is axiomatic that all fill sites incorporating domestic refuse or degradable organic materials will be subject to the hazards of landfill gas.

12.11.2 Constituents of landfill gas

Landfill gas is a variable mixture of gases, mainly methane and carbon dioxide. Under steady anaerobic conditions typical constituents and proportions at the seat of production are:

Methane	55–65%
Carbon dioxide	40–35%
Carbon monoxide	
Ethylene	
Hydrogen	
Hydrogen sulphide	
Mercaptans	remaining 1–5%
Nitrogen	
Oxygen	
Water vapour	
Trace gases.	

Figure 16 illustrates the various phases and variation in composition of landfill gas generation.

At certain stages of the decay the gas may be almost entirely carbon dioxide with methane practically absent. The presence of carbon dioxide in normal atmospheric air is very small (0.03%). Greater amounts are warnings of biodegradation, burning or other hazards. Under other special conditions the proportion of hydrogen or hydrogen sulphide may greatly increase, in the case of hydrogen up to 20%. This is normal for the anaerobic, non-methanogenic phase.

12.11.3 Generation of landfill gas

Present-day wastefill, of high putrescible content (see Figure 6, page 39), could theoretically generate over the whole period of biodegradation up to 400 m³ of gas per tonne of dry fill. In practice only 50% appears as gas: the rest is lost as leachate. Earlier wastefills, richer in ash, will generate substantially less gas before exhaustion. Wastefills over 30 years old are less likely to pose a gas threat to building works provided the degradation processes have progressed effectively during the period.

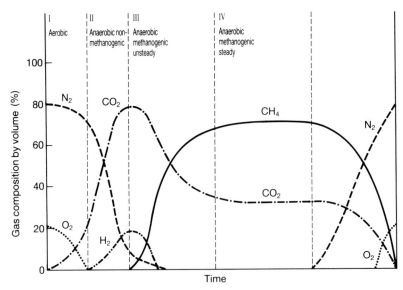

Figure 16 *Phases in landfill gas generation*

Methane-producing organisms function over a pH range of 6.6–7.6. If, during biodegradation, the rate of acid formation exceeds the rate of breakdown to methane, the pH decreases, methane production falls off, and the carbon dioxide content of the gas increases (Jarrett, 1979).

Under optimum conditions of pH = 7.0, temperature about 35–38°C for mesophyllic bacteria or 55°C for thermophilic bacteria (Jarrett, 1979), and compaction just adequate to ensure anaerobic action throughout, the putrescible content of modern wastefill may be largely biodegraded within 5–6 years, but paper and other less biodegradable materials may continue to decay for 30 years or so. More usually, under less favourable conditions, decay may be so retarded that the putrescible matter is still producing gas even after 30 or 40 years. Alternatively, if disturbed, it may revert to gassing after years of inactivity.

Toxic substances including heavy metals, if present in concentrations sufficient to affect the anaerobic bacteria, may seriously retard biodegradation.

For domestic waste the optimum moisture content for biodegradation is generally 40–60% (Jarrett, 1979), compared with the optimum for compaction of 50–70% (Harris, 1979). Wastefills, typically deposited at 20–50% moisture content, are thus on the dry side of optimum for both compaction and biodegradation and have a slow biological stabilisation rate. At higher water contents biodegradation is retarded. Gas can still be produced, even under water.

Compaction, whether under an imposed building load or resulting from dynamic operations, may increase the gas-generation rate through pushing putrescible matter towards zones of higher moisture content. In addition, short-term lateral migration will increase by the squeezing out of gas. Settlement of deep fill under its own weight may be expected to have a similar effect. Similarly, a rise in the water table may also result in an increased emission rate, because of the pumping action of the water.

Barry (1987) gives much useful information on the generation and behaviour of landfill gas.

12.11.4 Migration, emission and accumulation of landfill gas

Landfill gas can migrate considerable distances from its source following zones of relatively high permeability, including:

- Granular strata
- Fissures in clay and rock
- Cracks and joints in concrete
- Services, ducts, sewers and drains
- Voids round piling
- Vibrated stone columns.

Migration of landfill gas from an underground source is either a convection or a diffusion process. The presence of low-permeability barriers (clay, membrane, frozen or waterlogged ground, concrete slabs, etc.) may cause an increase in pressure resulting in greater migration.

The migration of landfill gas in the ground is affected particularly by ground permeability. Changes in porosity, caused by compaction, ground loading, application of impermeable layers, concrete slabs, etc. or changes in the water table, during the course of the works, will influence the migration rate and pattern. Gas migration is also affected by such factors as site geometry and decomposition history, which are less likely to change during site work.

Rapid increases in gas emission have been observed at the end of winter. This is probably due to the increased permeability of the ground (less waterlogging and frost effect) at the onset of spring. Ambient air temperature effects do not extend to great depth.

Emission rate is known to increase under falling barometric pressure. At such times there is an increased risk of build-up of gas in excavations, sewers and other cavities.

Gas emission discovered in a newly drilled borehole may be the result of continuing generation, or may be an accumulation in an isolated pocket which, once pierced, will soon disperse. Monitoring of the emission rate will normally distinguish between the two causes.

Landfill gas may be lighter or heavier than air, according to the proportion of methane to carbon dioxide. The mixture will be lighter than air when the methane content is about 55% or greater. Gas may thus accumulate at any level in an enclosed space and, in certain circumstances, will tend to remain in a layer. In still air, a layer of explosive and/or asphyxiant gas may remain for long periods at a given level without mixing other than by thermal diffusion.

There have been explosions caused by accumulation of landfill gas in buildings. These have been attributed to the lateral migration of landfill gas from old wastefill sites to the greenfield sites on which the houses were erected.

12.12 Aggressive attack and corrosion

A number of the contaminants found on derelict sites have the potential to attack building materials buried in or laid on the ground. In cases concerning the durability of various elements of the structure, reference may usefully be made to CIRIA Report 98 (Barry, 1983), which deals with the various mechanisms of aggressive attack in the ground. The potential for attack depends on:

- The availability of a given contaminant
- Contact between the contaminant and building materials
- Sensitivity of each material to the contaminant.

The aggressivity of a substance depends on its chemical condition in the ground. Materials which are attacked by aqueous solutions (such as sulphates) may be relatively unharmed by the anhydrous chemical. Deep drains, or pumping, to keep the water table permanently at a low level, may be an important protection. However, even small amounts of soil moisture may be sufficient to initiate attack.

Significant corrosion requires continuing contact of the chemical with the target material. With water-soluble reagents this implies upward or lateral movement of the groundwater (or downward percolation of surface-contaminated water) to replenish the depleting chemical. On the other hand, chloride can migrate and promote corrosion reaction without being consumed. Alternatively, evaporation from walls and other exposed surfaces may lead to continuing replenishment and increasing concentration through capillary migration of moisture.

The susceptibility of common building materials to attack by aggressive chemicals is discussed in Section 17.6.

12.13 Leachate and contamination of groundwater

On all types of derelict site the possibility of leachate production, and consequent contamination of surface and groundwaters and the aquifers, must always be borne in mind. Even though site building works may not be affected, these hazards may pose a threat elsewhere on or off the site. Furthermore, there will be statutory obligations regarding the protection of water resources and the environment.

Annual rainfall in the UK generally exceeds evapotranspiration, resulting in net infiltration. Percolating or migrating groundwater will become contaminated in contact with biodegradable wastes, leachable industrial contaminants, building wastes or other soluble or reactive materials. Once such contaminated groundwater migrates laterally away, or percolates downwards, into cleaner zones as a plume detached from the source contaminant, it becomes leachate.

Leachate may slowly release considerable pollution over a long period if it is allowed to enter surface or groundwater flows or the aquifer. Even though the aquifer may be a considerably larger volume of water, there may be little dilution of the plume since movements are extremely slow, and aquifer and plume tend to laminate rather than mix. Dangerous mixing may occur, however, when conditions are disturbed (for example, by attempts to extract a fresh water supply by pumping from a well in the 'laminated' zone). On the other hand, lateral migration in sloping terrain may cause the leachate to emerge in lower-lying surface ponds, streams, etc.

Regarding leachates from biodegradable wastes, young fills (less than 20 years old) are generally the more troublesome. In older wastefills anaerobic degradation should have reduced the greater part of the putrescible materials to simpler components (CH_4, CO_2, N_2, etc.). Table 11 illustrates the striking difference in leachate from young and old fills.

Table 11 Composition of typical landfill leachates
(after Young and Barber, 1979)

	Fresh refuse	Aged refuse
pH	7.1	7.1
Chemical oxygen demand (COD)	11 600	125
Biochemical oxygen demand (BOD)	7 150	5
Total organic carbon (TOC)	4 440	40
Cl	2 100	225
SO_4	460	—
Na	2 400	240
Mg	390	90
K	490	75
Ca	1 150	280
Fe	160	0.1

Values in mg/l except pH.

Industrial sites can also produce hazardous leachate. Frequently they develop a layer of contaminated groundwater, which may or may not be in motion away from the site. Demolition of buildings or digging of trenches can alter the hydrological regime.

Solvents and other organics, plating liquors, etc. are frequent contaminants. Other hazards may be acid run-off from colliery spoils and sulphur-rich leachate from slags. Organic liquids, immiscible with water, may float on the groundwater or sink below it until they reach a relatively less permeable stratum, where they move independently of the groundwater. They may flow in a different direction, depending on the surface topography of the low-permeability layer.

The rate of production and movement of leachates under most conditions is very slow. Considerable time may elapse before the effects of a change in the regime become apparent.

Another circumstance, particularly relevant in large towns, which concerns the leachate hazard is the rise in groundwater levels. In large cities such as London, Birmingham and Liverpool these groundwater regimes have changed by reduced pumping for water supply and increased leakage from mains. Deep aquifers and perched water tables have been affected. In such places rising water may bring underlying contaminants towards the surface.

12.14 Migration of contaminants

An important phenomenon in the behaviour of a site containing contamination is the long-term migration of the contaminant itself, generally by movement of water-soluble substances through the adjacent natural ground. Even where the contaminant has been encapsulated within artificial barrier walls, etc., failure of the encapsulation can lead to unwelcome migration. Thus failure of a cover layer may permit upward migration, which can put at risk the topsoil or other sensitive targets at ground level. Failure of an underground barrier wall may permit lateral migration, with consequent threat to adjacent property. Permeable strata below the contaminant may allow downward migration, especially where surface water is able to percolate, leading to possible contamination of the aquifer (Figure 17).

Figure 17 *Contamination seeping through brick lining of a tunnel (photograph by courtesy of GLC)*

Upward migration may occur through fluctuations of the water table or through soil suction on the drying out of fine-grained upper soil horizons. Another cause may sometimes be the squeezing upward of thick, oily liquid contaminants (tars, etc.) under traffic or settlement of overlying fill.

Soil suction can be a serious cause of upward migration. Under drought conditions dried-out clay, for example, can exert sufficient capillary suction to raise soluble heavy metal salts for heights of 3–4 m above the water table. (See also Section 19.1.)

Doubts have been expressed as to whether heavy metals do, in fact, migrate upwards, unaided, under the influence of soil suction alone. Elevated levels of heavy metals in topsoils have been considered explicable by other mechanisms:

- Airborne dust pollution from smelters, mines, etc.
- Metals already present in supposedly clean topsoils
- Naturally higher levels in the normal topsoil derived from the decomposition of the local bedrock
- Uptake by deep-rooted vegetation.

Such factors are doubtless of importance and should not be overlooked. However, carefully controlled experiments (Cairney, 1985; Smith and Bell, 1985) have proved beyond doubt that heavy metals do migrate upwards unaided, and have established the soil suction conditions governing the rate of such migration.

12.15 Biodegradation of refuse fill

A general understanding of the processes of biodegradation may be useful in the interpretation of conditions (gaseous emissions, leachate composition, etc.) brought to light upon opening up, or sinking boreholes into, ground containing refuse. It is important to appreciate that the composition of gases emitted may change within a few hours or days of opening up if anaerobic action reverts to aerobic. (See also Sections 12.11 on landfill gas and 21.4 on leachate monitoring.)

Within the unsaturated zone of a refuse fill, leachate formation may start with the hydrolysis and aerobic oxidation of refuse, resulting in the conversion of insoluble solids into water-soluble substances and complex organic compounds into simpler components (Naylor *et al.*, 1978). In zones of oxygen shortage, anaerobic degradation of carbohydrates and reduction of sulphates and nitrates then yield carbon dioxide, methane, ammonia, hydrogen sulphide, nitrogen and other simple substances.

During biodegradation the carbon content of the fill drops. Loss of mass under controlled combustion has, therefore, been suggested as an approximate yardstick for the potential for further decomposition (Harris, 1979). However, the reduction of carbohydrates into methane and carbon dioxide is very slow, and the lower carboxylic acids (e.g. acetic, propionic, butyric), generated as intermediate products, form a high proportion of the organic content of the leachate.

During biodegradation metals tend to become attenuated by precipitation, either as carbonates or hydroxides during aerobic action or as sulphides during anaerobic action (Naylor *et al.*, 1978). Organic decomposition processes slow radically within a few years, whereas the low-solubility metal compounds continue leaching for an indefinite period. Landfill leachate tends to become rich in minerals (Jarrett, 1979).

The foregoing applies particularly in the case of plain refuse deposits of putrescible material. Other materials, such as paper, as well as plastics and other hydrocarbon-based products, are much less easily biodegraded. In the somewhat less common case of co-disposal of industrial and domestic waste the chemistry is different. This has an influence on the component substances to be detected during monitoring of the leachate (see Section 21.4).

13. Constraints on development

This section discusses in more detail the basic concepts of constraints outlined in Section 4, under the following related headings:

- Technical and environmental
- Statutory
- Safety
- Time and delays
- Cost.

13.1 Technical and environmental

13.1.1 Relationships between hazards and end use of site

Constraints arise from the interaction between the hazards on a derelict site and the end use proposed for that site. Many complex factors affect that relationship, but experience gained over the past decade or so is now sufficient to permit at least a qualitative assessment.

Table 2 (inside back cover) is an attempt to set out in broad terms this fundamental complex of relationships for the more common types of derelict site and hazards, and for the targets implicit in different end uses. (The table is explained in Section 3.2.)

13.1.2 Gaps in present knowledge

At present, sufficient quantitative data are not yet available to permit an explicit, analytical appraisal of many essential aspects of derelict land, including:

- Sensitivity of end use to specific contaminants
- Effects of aggressive chemicals on buildings, foundations and services
- Effectiveness of remedial measures against aggressive chemicals
- Factors influencing potential combustibility
- Generation and emission of landfill gas
- *In-situ* treatments of contamination
- Movement and quality of leachate.

Engineering decisions which involve these aspects will therefore call for a great deal of judgement, relying upon the engineer's own practical experience as well as that of expert advisors. Research is in progress worldwide on these and related factors. Smith (1985) gives a state-of-the-art review of available reclamation technologies and their probable effectiveness. A second study has started (Smith, 1987b) and is concerned with practical evaluation of a range of technologies in a series of demonstration projects for clean-up technology and protection of groundwater.

13.1.3 Basic technical constraints

Inadvisable developments. From a study of the present state of the art, certain basic concepts emerge as to what is considered good practice today. Unless adequate provisions (compaction, piling, covering, etc.) are made, buildings should not normally be located:

- Across a landfill/natural ground interface (differential settlement risk)
- On actively gassing sites and domestic refuse fills less than 20 years old (excessive settlement and gas)
- On loose fill not known with certainty to have undergone collapse settlement (rapid settlement risk)
- On fill liable to spontaneous ignition
- On loose fill of rapidly varying depth (differential settlements)
- In situations where groundwater migrations can bring aggressive chemicals into contact with building foundations, services, etc. (corrosion, health hazards, etc.)
- On sites of old coal-washery lagoons (excessive settlement and combustion risk)
- On sites impregnated by recent tar or thick oil spills (settlement by squeezing out fluid)

- On sites with free sulphur or sulphide, or even calcium sulphate (H_2S generation, sulphate attack, expansivity hazard)
- On sites with unacceptable substances present. Asbestos as airborne dust, dioxin, and radioactive substances, *inter alia*, are considered to be unacceptable.

Safety and containment of contamination. A contractor on a contaminated site has not the freedom of working to be expected on a greenfield one. From the outset, all site operations should be planned to avoid any risk of spreading local concentrations of contamination onto hitherto clean areas on- or off-site.

Wherever possible, there is an advantage in buying land for reclamation with all plant, etc. still intact and standing. Demolition should be in accordance with BS6187: 1982, *Code of Practice on demolition*. Where such plant has already been dismantled, usually at minimum cost for the value of the scrap, there is a high risk of finding the site damaged by careless spreading of contaminants.

Examples of bad practice (Smith, 1982) have included bulldozing of stockpiles of spent oxide across clean areas of a gasworks site; failure to remove drums of hazardous materials, resulting in spillage on an otherwise clean site and risk to demolition workers; flooding of tar over extensive clean areas through careless demolition of a coking plant (Figure 18).

Figure 18 *Flooding of tar over clean area due to careless demolition (photograph by courtesy of M. A. Smith)*

Against such contingencies the contractor may be obliged to provide temporary tanks for the emergency storage of dangerous liquids. Contaminated water may require dilution to an acceptable level before a local authority will consent to its discharge into a foul sewer. Great care must be taken during drilling or pile driving through underground tanks, pipes, etc., and safety expedients (bentonite sealing, etc.) must be ready to prevent the release of dangerous materials into the aquifer or clean ground. Specialist contractors may be required for the removal of impregnated ground to an appropriately licensed tip before the main works can start.

Onerous expedients may be required to prevent the escape of harmful contamination and other hazards beyond the confines of the site. These may include measures against airborne dust, gas and troublesome odours, migrating harmful leachate, gas and fire on or under the ground, as well as traffic-borne contamination on the wheels and bodies of vehicles.

The protection of site workers under the Health and Safety at Work Act 1974 will require additional steps in the programming of site works, as well as special safety procedures, facilities, equipment and clothing.

13.2 Statutory

The Environmental Protection Act 1990, contain provisions of direct relevance to gas-emitting and contaminated sites, to the disposal or re-deposition of wastes, and to the duties and powers of local and waste regulation authorities.

13.2.1 Planning control and the relevant authority

Existing legislation confers wide-ranging powers on the planning authority in the control of development of derelict sites. At the same time, the planning authority has the obligation to satisfy itself as to the continuing safety of a proposed development, site workers, future occupants and third parties. The statutory constraints applying to derelict land are thus considerable. In addition to the Building Regulations 1985, legislation on planning, waste disposal, pollution control and health and safety imposes further constraints on development.

The DoE has issued three relevant circulars:

- The use of conditions in planning permissions (Circular 1/85 of 7 January 1985)
- Development of contaminated land (Circular 21/87 of 17 August 1987)
- Landfill Sites: Development Control (Circular 17/89 of 26 July 1989).

These are guidelines to the planning authorities on the exercise of their considerable powers and responsibilities. These documents should be carefully studied along with DoE Planning Policy Guidance PPG 14 *Development on unstable land* (HMSO, 1990).

For the Acts, Orders and Regulations relevant to building on derelict land in England and Wales reference should be made to the latest edition of CIRIA Special Publication 23, *Building design legislation* (CIRIA, 1982) or similarly for Scotland to CIRIA Special Publication 34, *Scottish building legislation* (CIRIA, 1985) as well as Appendix 2. (These publications also indicate legislation relevant to building in general.) The major role in the implementation of these extensive regulations is played by the local authority. The attitudes of such authorities and their interpretation of the legislation can play an important part in the development or otherwise of a derelict site.

Other authorities will be concerned as to the enduring safety at all times of all services (water, gas, electricity, telephones, etc.) on the site or nearby and can insist on certain works being carried out. Ancient monuments and remains and sites of special scientific interest must likewise be safeguarded or dealt with as prescribed by the interested authority.

The National Rivers Authority has statutory control over the quality of discharges to drainage systems and water courses and is also charged with preservation of quality of water supply.

Of equal concern are proposals for coping with unforeseeable emergencies during both the site investigations and the development itself. In high-risk situations, with the possibility of fire or explosion or involving removal of highly toxic or radioactive materials, the police, fire and medical services will be involved and can impose constraints. In the special case of radioactive contamination the National Radiological Protection Board, Rowstock, Oxon, will be involved.

13.2.2 Attitudes of statutory authorities

It is generally in the interests of a local authority to permit development of a derelict site. However, the authority is legally bound to require satisfaction that the development will be without hazard to the public at large, site workers and future users of the development. On the other hand, the granting of planning permission implies no guarantee on the part of the authority that the site is safe or suitable for the proposed development. Responsibility for its safety rests with the developer.

The engineer should ascertain, at the outset of the initial assessment, the attitude of the local authority to a proposed development. Attitudes vary across the country to such an extent that a particular type of development might be forbidden in one locality, permitted without hindrance in another, or allowed to proceed only after a formal 'waiver' in yet another.

13.2.3 Interpretation of regulations

In consequence of the present imperfect knowledge of the real effects of many hazards and the efficacy of remedial measures, statutory authorities may vary in the stringency of their interpretation of the regulations relating to derelict land. The following are typical examples of some of the constraints resulting from statutory regulations.

Gas generation. The Public Health Act 1936, Section 54, imposed a duty on a local authority to forbid building on ground containing 'offensive material' unless removed or otherwise rendered innocuous. With wastefills some authorities took the view that the presence of landfill gas, no matter what the rate of emission, was evidence of offensive material (Durn, 1983). If gas was detected, Building Regulation permission was refused. Although the above Act has been superseded by the Building Act 1984, such attitudes may persist.

Leachate. Water authorities accept that leachate will form and must be taken into account, but require stringent protective measures where ground or surface water or the main aquifer are at risk. The usual remedy on building development sites calls for additional works to reduce the volume of leachate generated and to control its migration. Special alternative processes also exist for treating the toxic content of the leachate. These pertain more particularly to waste management and are probably less likely to be required of a developer.

However, the NRA or water company may demand acceptable dilution or treatment of contaminated water prior to discharge into a foul sewer or other watercourse, or that it is disposed of as a special waste.

Water supply. Under the Model By-laws, water authorities do not permit the laying of water mains until their own tests have shown the ground to be acceptably free of contamination. Hitherto the practice in laying service pipes, etc. in contaminated ground has been to excavate large trenches to permit surrounding the services in clean material, sufficiently oversized to allow maintenance without disturbance of the surrounding contamination.

Some water authorities no longer accept this expedient, which is considered to form a drainage path for contaminants (Durn, 1983), with the possibility that a burst may result in back-siphonage into the main. Instead, services are today generally required to be laid in a cover layer (see Section 19.1) above the contaminant.

Combustibility of tips. Tests are required to assess the potential for outbreak of fire in case a tip is opened up, for site investigation, etc. However, to date, no agreed standard has been established for assessing combustibility potential, and authorities vary in the stringency of tests demanded.

Waste transport. Recent legislation requires registration of waste disposal transport companies in order to curb fly tipping of site arisings.

13.2.4 Waiver of by-laws

Some local authorities may adopt a flexible interpretation of the regulations, but insist on full public safeguards. In the past this has sometimes required a waiver to a specific regulation. Such a waiver would be granted only if there were sound evidence for the efficacy of the measures proposed.

A somewhat similar procedure operates today. Under the Building Regulations 1985, Approved Document C lists recommended procedures for dealing with possible contaminants. Local authority approval is required for measures other than those recommended. Approval requires satisfactory demonstration of the effectiveness of any alternative proposal.

13.3 Safety

The need for safety is paramount in all phases of a derelict site development. It imposes continuing constraints which must be taken into account in determining the economic feasibility and the programming of the development.

The developer has full responsibility for the overall safety of the finished works, future occupants and third parties. The contractor has responsibility for the safety of site working and third parties thereby affected. The contractor's safety measures (Section 23) may be onerous and will influence the contract price. Furthermore, negligence by the contractor regarding safety may involve the developer and the engineer in litigation.

Where dangerous material is to be removed from the site the developer's responsibility also persists off-site. With 'special wastes' this responsibility will extend as far as an appropriately licensed tip, which may be at considerable distance, and costs will be correspondingly high.

Where a site is not to be developed, the owner has responsibility for the safety of third parties and the environment. This may entail engineering works (Section 7.4) and corresponding costs.

13.4 Time

Planning. Experience has shown frequently that insufficient time is dedicated to the proper appraisal of a derelict site and the drawing up of a rational development plan (Lowe, 1984). One solution might be the prior execution, by a local or regional authority, of a broad overall appraisal of potential problems over a large area or district, as opposed to a specific site. A good example of one such broad appraisal, or prospective survey, is the *Survey of Contaminated Land in Wales* (Welsh Office, 1988). To date, however, prospective surveys exist for only a small fraction of the derelict land in Britain, and it is unlikely that they will ever be available in time for the detailed development of the majority of derelict sites.

In the absence of a prospective survey the Environmental Health Officer is usually the individual most likely to be aware of potential problems in a derelict site. Current planning application/consent procedure puts pressures on Environmental Health Officers. Under present UK legislation, as the public safeguard on derelict land, they have a duty to veto any proposal in which there is uncertainty regarding environmental safety.

By consulting the local authority's Chief Environmental Health Officer in advance of a planning application a developer can do much to alleviate such pressure, and so avoid the situation in which environmental hazards come to light so late as to delay planning decisions. This also underlines the importance of the submission, by the developer with the planning application, of a properly executed record of the past history of the site.

Site works. Site investigation may be expected to take considerably longer than for a greenfield site, especially where problems of gas emission, fill settlement, leachate or contaminant migration, etc. may require monitoring over periods of at least six months.

During the reclamation and construction phases safety procedures and ground treatments will be time consuming. Working procedures will also be prolonged. For example, densification of cohesive fills may require extra time for dissipation of pore water pressures (Section 18.5.3), while massive obstructions may need excavation and breaking out prior to piling.

Furthermore, even with a detailed site investigation, hazards may sometimes only come to light when the site is opened up for construction (see Section 5.3). Time should be allowed for dealing with such unforeseen eventualities, including possibly a complete re-planning of the development. The development programme should thus allow considerably more time for its execution than with a greenfield site.

13.5 Cost

The consequence of these various constraints is that the works required for the development of a derelict site will prove more costly than would be the case on a greenfield site.

However, given the current high cost of purchasing greenfield land, such extra costs on a derelict site may be justifiably set against the enhanced value of the land resulting from the remedial works.

In addition, the political urge towards the re-use of derelict land has resulted in public money being made available in some areas (see Appendix 1) to local authorities and to developers towards offsetting the extra costs.

14. Initial assessment

Initial assessment is the first phase of investigation strategy (Section 5) and involves desk study and site reconnaissance, usually a multi-disciplinary exercise. Its purpose is a preliminary appraisal of development potential, but full appraisal will require the further investigations outlined in Section 15.

All the stages of investigation should conform to the recommendations of DD175 (BSI, 1988) and BS5930 (BSI, 1981).

14.1 Multi-disciplinary approach

The appraisal of a derelict site and the choice of remedial measures require specialist skills in support of ground engineering technology. In particular, contaminated sites require:

- Identification of chemical, biochemical and radiochemical contaminants and their concentrations in the ground
- Evaluation of the effects of such concentrations on targets exposed by the proposed end use. Ideally, this would involve comparison with standard reference values of acceptable concentrations but, even where such reference values exist, their application requires expert judgement.

This demands an expertise and a knowledge of the behaviour of contaminants in the ground, for which the engineer will have to call upon the skills of suitably experienced scientists, e.g. chemist, microbiologist, radiochemical specialist, etc. A botanist could also be needed even though plant life might not appear to be affected by contamination. Similarly, in order to assess the migration of waterborne contaminants through the ground a knowledge of local hydrogeology will be essential. Other workers, whose collaboration will be valuable, include persons experienced in the process previously operated on the site, and with a knowledge of the layout of the plant involved.

If the engineer does not have the relevant geotechnical expertise a suitably qualified and experienced geotechnical engineer will have to be brought into the team at the start. As soon as contamination is suspected an appropriately experienced scientist should be added: often this will also be at the outset. This choice is vital, because the scientist's expertise will be needed throughout the reclamation and construction stages as well as in the appraisal. Knowledge of industrial applications and an understanding of the environmental implications are essential criteria for the selection of the scientist.

It is rarely necessary to involve other experts at the outset, but they should be called in as soon as the need for their services is indicated by the investigation work.

The problems raised by derelict sites are so interactive that calling for advice on specialist aspects on an *ad hoc* basis is rarely satisfactory. The team is required to design and specify the detailed investigation and the ensuing remedial works and therefore needs to be fully conversant with all relevant facts at all times. Knowledgeable co-ordination is essential, and thus the employment of separate organisations for different aspects of the work is not recommended.

The client's best interests will be served by the establishment of a consultancy team on a professional basis to specify and control the fieldwork. Alternatively, the client can appoint an organisation, responsible to the engineer, which embraces all the necessary disciplines.

It should always be recognised that, although in many ways the investigation work is similar to that of conventional site investigation, derelict sites demand much additional experience and knowledge which is currently available only from a limited number of consultancies and specialist organisations.

14.2 Desk study

14.2.1 Information required for all sites

Dumbleton (1979) and other sources summarise the required information thus:

- Nature of site; details of present and past owners or occupiers
- Layout of site at various stages in its development; buildings, surface and underground installations; extent of persisting remains
- Activities and processes carried out and their locations; raw materials and products; waste products and methods of disposal
- Nature and extent of any contamination of the ground, groundwater or of remaining installations
- Extent, composition and age of fills and other deposited materials, and of any stored materials, including those in buildings
- Relevant information on site geology, soils, meteorology and hydrology, sufficient to distinguish natural from artificial conditions.

Information compiled from the desk study needs careful correlation and reappraisal with the observations made during the field reconnaissance of the site. If sufficient information on past use appears to be unobtainable from a reasonably extensive search of available sources, appropriate additional studies should be allowed for in the detailed site investigation.

Historical site surveys are referred to in BS5930: 1981, *Code of Practice for site investigations*. Although intended for engineering investigations of greenfield sites, the Code contains much that is relevant to contamination surveys.

14.2.2 Previous use of the site

One aim of the desk study is to identify previous usage of the site and types of process carried out, and hence the types of contamination or other hazard to be expected. A gas works, for example, may have recognisable production equipment and storage areas with related chemical products and by-products. Plans may be available showing the layout of each component of the works. Other works may not be so well documented. The study of a similar works elsewhere may then be useful.

A site may have had more than one usage in the past. Scrapyards are an obvious example; infilled dockland may reveal contaminated silt covered with all types of industrial waste; gas works may have been built in the last century on the site of earlier industries. Similarly, nineteenth-century houses were often built on previous industrial sites, or over dwellings where cottage industries once thrived. Successive rebuilding may have taken place over earlier foundations, so that plans made just prior to works' closure may not reveal all foundations and buried tanks.

The recent reclamation of Thamesmead (Lowe, 1984) over a site of 400 ha revealed a complex of munition factories, gas works, power stations, etc., with railroads and sidings built on sites formed of wastes of earlier brass foundries and other industries dating from the seventeenth century and marshland 'reclaimed' with waste in the nineteenth century.

14.2.3 Additional information for wastefill sites

Particular care is needed regarding the documentation of wastefill tipping. Records may be scarce or non-existent (especially before the Deposit of Poisonous Wastes Act 1972), even where the local authority may have been the tip owner. Information of some value may sometimes be obtained from local inhabitants and past or present employees of the authority or from works adjacent to the tipping site.

The absence of information on the wastes present within a fill will increase the need for an exploratory drilling and sampling campaign (see Section 15).

For wastefill sites, in addition to the type of fill materials, the age of the deposit is also important, as it has a bearing on the ash content and degree of biodegradation of the fill—both important factors in assessing the potential for settlement and methane generation.

Information on the methods of disposal (dispersal on surface, lagooning, mixing, etc.) and the locations of various wastes in the fill, including depths of fill material, will prove useful and can simplify the form of any subsequent investigation. The positions of lagoons used for liquid disposal may be evident from a site inspection or located from aerial photograph records (Figure 19).

Figure 19 *Excavation of lime pit in former chemical works (photograph by courtesy of Thorburn Associates)*

Records of previous monitoring of wastefill settlement may be scarce but can provide important historical evidence. Where time permits, settlement measurements over a sufficient period by careful levelling from a reference datum together with recorded water table levels may give a valuable indication as to the stage of settlement reached. Data from on-going methane or groundwater monitoring may be available and may permit an inference as to nature and condition of the fill.

14.2.4 Historical evidence

Historical evidence is available in widely differing forms. Local historians or archivists can greatly assist in obtaining such information. The evidence may include:

- Topographical and special maps
- Publications on local history or industrial archaeology
- Oral evidence from previous occupiers of the land and their employees, as well as local inhabitants
- Technical process and waste-disposal records
- Business records, old maps and plans, directories and rating lists
- Previous site-investigation records
- Settlement-monitoring records
- Site development and construction records; detailed layouts and drawings, especially those showing deep foundations, buried tanks, services and other components of the works
- Early proceedings of the Institution of Civil Engineers and other relevant professional bodies. These can provide drawings and records from before the first edition large-scale Ordnance Survey maps
- Aerial and other photographs of the site and its environs
- Local historical records.

Aerial photographs are usually available going back to about 1945 and occasionally earlier. Guidance on obtaining, examining and interpreting such photographs has been given by Dumbleton and West (1970, 1976) and on the particular application to derelict land by Bush and Collins (1972).

Vertical aerial photographs should be studied stereoscopically and should therefore be purchased with adequate overlap between adjacent frames. Stereo photographs provide a wealth of detailed information not otherwise shown on drawings. Historical series of photographs, where available, can be used to follow the evolution of a site and the chronology of filling operations.

14.2.5 Sources of information

Sources of information may include:

- Archives from industry, museums and libraries in the locality
- Relevant departments of local authorities (county and regional authorities, boroughs and district councils) and academic institutions
- Factory and Alkali Inspectorates
- Local authority poisonous waste units
- National Rivers Authority
- Refuse-disposal tip operators
- Aerial photographs held by the Ordnance Survey, Royal Air Force, the Royal Commission on Historical Monuments and various other commercial organisations.

Table 12 summarises general sources of information. Reference should also be made to the detailed list given in the BS Draft for Development DD175 *Code of Practice for the identification of potentially contaminated land and its investigation* (BSI, 1988).

Table 12 Summary of sources of information for the appraisal of derelict sites

Topographical maps, plans, and charts
- Ordnance Survey maps, town maps, insurance maps, deposited plans (Parliamentary plans for railways and canals; building permission plans)
- Estate maps, enclosure maps, tithe maps; Admiralty charts

Special maps and related literature
- Geological maps, soils maps, land-use maps; accompanying memoirs and literature; advice of publishing bodies

Meteorological and hydrological records

Air and ground-based photographs; pictorial views

Local and topographical literature
- Victoria County History, local histories, local newspapers
- Company histories, house magazines
- Proceedings and publications of local history, archaeology and industrial archaeology societies

Directories
- Local street and trade directories, directories of individual trades and industries

Technical and professional literature
- Histories and technical literature of individual industries; proceedings of professional and technical institutions

Owners and occupiers
- Information and records on site investigations, site development and plant, technical processes; site plans; business records

Local authorities
- Advice and records; rating return and valuation lists; derelict land units; waste-disposal authority

Service authorities
- Records of buried services: water, sewers, gas, electricity, telephone

Records of mining, quarrying, opencast extraction

Factory Inspectorate; Alkali and Clean Air Inspectorate

With regard to leachate-producing sites the WRc Technical Report TR91 (Naylor *et al.*, 1978) gives sources of statistics and other data on local meteorology, evapotranspiration, geography, geology, hydrogeology and other natural factors affecting leachate production, movement and attenuation.

Figure 20 *Excavations covered by a car park (photograph by courtesy of Environmental Safety Centre, Harwell Laboratory)*

14.3 Site reconnaissance

14.3.1 Scope

Site reconnaissance provides visual evidence to confirm and supplement the information obtained from the desk study. In addition to the objectives for a greenfield site, reconnaissance of a derelict site is normally concerned with the discovery of contamination and obstructions as first priority. Geotechnical characteristics will be the subject of qualitative examination. It will be beneficial for monitoring of fill sites to be initiated at this stage.

A contaminated site should be thoroughly inspected before demolition or clearing if possible. At least one visit should be on a day of heavy rainfall to obtain evidence of possible waterborne pollution, overland contaminant flow, and surface erosion. A site still occupied by a functioning works should be visited to profit by discussions with management and workers (Figure 20).

The reconnaissance should be carried out by the principal member(s) of the engineer's team. As a minimum, this should be the geotechnical engineer who, if contamination is suspected, should be accompanied by the environmental scientist.

Reconnaissance of a contaminated site will expose the engineer and other members of the team to potential health hazards. It is not usual practice to install full safety facilities on site for this exercise, but the precautions set out in Section 23 should be implemented as appropriate.

14.3.2 Direct evidence

One or more of the following hindrances to development may be in evidence on the site. All should be inspected, photographed and carefully recorded on drawings (Dumbleton, 1979):

- Extant buildings and other structures, including tanks and pipework
- Hard-covered areas of ground, such as roadways, railway sidings, storage areas, vehicle parks
- Domestic/commercial waste-disposal fill; inert opencast fill, other fill

- Abandoned pits, depressions, quarries, lagoons, ponds with or without standing water
- Made ground, mounds, waste heaps
- Sewers, manhole and tank covers; old petrol pumps
- Drums, barrels, kegs which may contain hazardous substances
- Signs of settlement, subsidence, filling, disturbed ground
- Material such as sewage sludge, slag, ashes, asbestos-lagging scrap, and industrial or chemical waste at or near the ground surface
- Discoloured soil, blighted vegetation, significant odours
- Polluted water and other fluids in seeps, puddles, ponds or streams.

Samples of readily accessible materials from the surface soil, surface water and ambient air should be collected for further examination and/or analysis. The procedures set out in Section 16 for collection and storage of samples apply also to this initial site reconnaissance.

14.3.3 Indications requiring expert interpretation

There may be, in addition to the above, several indications more readily detected and interpreted by experts. Further visits by specialists may be required to complete the initial assessment. On the initial site visit the engineer should look for evidence that might require specialist interpretation. Some specific examples are outlined below.

Underground structures. Gantry crane and building columns and heavy machinery often have deep, massive foundation blocks, of which only a reduced plan area may be visible at ground level. Gas works retort houses, etc. have deep basement structures, often filled in on a derelict site and no longer evident. These structures, underground pipe runs, tanks, etc. are more easily interpreted and located by experts from the industry concerned.

Surface materials. Unusual colours and contours may betray identifiable chemical wastes and residues.

Fumes and odours. A number of substances are readily detectable and identifiable at very low concentrations.

Landfill gas. Suspected deposits of putrescible matter should be recorded for later sampling.

Flammable or toxic emissions may be detected with portable instruments: hot-wire katharometers; gas detector tubes; infra-red analysers. Wherever possible and practicable, checks should be made for gas by sampling in confined spaces, holes in the ground, burrows, drains, etc. as well as by narrow-bore (12 mm) sampling probes. Such methods are not a valid substitute for monitoring.

Monitoring for gas emission should start as early as possible (Section 21.5). It can also be advantageous to initiate a spiking survey at this stage. Appendix 4 gives guidance on the necessary procedures.

Vegetation. Poor growth or absence of vegetation may indicate phytotoxicity. Survival of a single species, or of a restricted range of species, may characterise phytotoxic mineral waste. Thus tolerant strains of the Common Bent-grass (*Agrostis tenuis*) in the absence of other vegetation may betray metal contamination. The absence of earthworms in the topsoil may also be significant.

Old sewage sludge may support luxuriant growth of a limited range of species resistant to heavy metals. Unhealthy vegetation on fill sites may be a sign of underground methane generation or an abnormally high water table.

Noise pollution. Industrial processes in the vicinity may continue to create noise of sufficient level to be disturbing to a proposed residential development, especially at night. By day, a casual site inspection may fail to detect such noise. Expert evaluation may involve sound surveys over the audible spectrum for a full 24 hours. Similar considerations apply to ground vibrations provoked by power hammers, drop forges, etc.

Airborne pollution. Fumes and dust from sources outside the development may be overlooked if intermittent, but may have serious nuisance value to a proposed development.

Their evaluation may require air sampling and monitoring with wind rosettes during the detailed site investigation.

Surface waterborne pollution. Pollution from upstream sources may be carried across the development site by streams. The source of pollution may be detected during the initial site reconnaissance.

Intermittent contamination. Intermittent overland flow of contaminants may occur at times of rainfall-excess.

Gully and sheet-erosion of fills.

Contamination from external, extant sources. This includes emissions from smelters, contaminating solutes in streams and overland flow.

Suspect combustible materials. On sites where combustible material is observed or suspected from the desk study, soil temperatures should be measured by means of *in-situ* probes as described in Section 21.6. Early monitoring of soil temperature should be initiated as soon as possible.

Radioactive contamination. If radioactivity is suspected from the desk study, the National Radiological Protection Board should be consulted and their advice followed regarding the content and execution of both the initial assessment and detailed site investigation.

14.4 Monitoring

Wherever practicable, it is advantageous to initiate early monitoring of relevant site conditions during the initial assessment works, and to continue this without interruption as long as possible during the subsequent phase of site investigation. Section 15.3 outlines aspects requiring monitoring, and Section 21 the requisite procedures.

14.5 Information for planning detailed site investigation

In contrast to the site reconnaissance, the detailed site investigation is likely to involve considerable site activity, with probably a specialist contractor's workforce, mechanical plant, site offices and other temporary buildings. These activities will require careful advance planning to ensure their safety, economy and efficiency.

In the initial assessment all possible information relevant to such planning should be gathered during the site reconnaissance, including:

- Access to working sites and adjacent land
- Obstructions such as: foundations, power cables, trenches, fences
- Area for offices, field laboratory, plant depot, sample storage (preferably upwind of any source of airborne contamination)
- Ownership of working sites
- Site water supply, location and quantity available
- Available power supply
- Local infrastructure: telephone, employment, fire, police, transport and other services
- Arrangements for containment of accidentally spilled/displaced liquids, or for disposal by pumping with local authority approval and for disposal of excavated wastes as necessary
- Need or otherwise for dust-suppression measures
- Arrangements for emergency action.

The information derived from the initial assessment must be sufficient to design and estimate the likely costs of a detailed site investigation.

14.6 Report on initial assessment

Only rarely will the information gained during the initial assessment be sufficient by itself to answer fully the fundamental question posed in Section 3:

> Is the condition of the site compatible with the proposed end use and if not, what measures are necessary (or possible) to reduce the incompatibility to acceptable proportions?

Nevertheless, the qualitative information acquired about the former uses and hazards of the site should be sufficient for a preliminary judgement as to the feasibility and probable cost of a desired development and advisable corrective measures.

Some site hazards, such as active gas generation, if suspected from the initial assessment, may immediately rule out any form of development or, at least, will require caution at this stage. In the latter case definitive decisions should await a full-scale site investigation, with monitoring as appropriate.

The engineer should prepare a full report on the initial assessment, the information so far gained and its implications, and the stage reached in the full appraisal of the development potential of the site. Details should also be included of the further full-scale site investigation considered desirable, with an estimate of cost. As far as is possible at this stage, the report should also assess such forms of development as appear feasible, their costs and technical feasibility as well as advisable remedial measures to improve the safety and suitability of the site. This assessment should not be based on individual hazards taken in isolation, since a combination of hazards may prove greater than the sum of the parts.

At this stage it is important not to over-react to hazards brought to light in the initial assessment. The more detailed site investigation may well show that the hazards are not widespread.

15. Detailed site investigation

15.1 Objective

Site investigation is intended to provide adequate quantitative data to permit full appraisal of the site. It demands accurate survey, sampling and analysis and on-site geotechnical testing. It should reveal the various physical and chemical hazards relevant to the proposed end use, and the location, extent and severity of each such hazard.

Although modest in comparison to the overall cost of a project, detailed site investigation is a costly operation, especially for contaminated sites, by reason of the chemical analyses of the samples. Nevertheless, it should always be carried out as thoroughly as possible because the consequential cost of failing to identify significant contamination in the investigation stage is likely to be much higher. To that end, the initial assessment should have been so planned and exploited that the resulting information is of value in defining a cost-effective programme of sampling and analysis for the detailed site investigation, as well as indicating potential hazards to personnel.

15.2 Standards and workmanship

Practices typical of the geotechnical investigation of greenfield sites are not adequate for derelict land investigations. The heterogeneous distribution of the various materials, and the consequent impossibility of interpolating between sampling points, demand careful attention in the design of the sampling pattern and tight survey control to ensure accurate three-dimensional location of each sample taken.

Care must be taken to avoid degradation of samples through mixing with soil, ground or surface water, drilling fluid or other extraneous matter. Exposed sampling points in trial pits or boreholes should not be subjected to cross-contamination by airborne dusts or surface water. Vehicles on the site should not be parked, fuelled or repaired near sampling points.

The protection of personnel is of paramount importance during all site operations. Where appropriate, the safety provisions of Section 23 should be observed at all times during site investigation.

15.3 Monitoring

The investigation can only yield information on the current state of a leachate plume, gas emission, settlement, etc.: it gives little indication of the likely changes in such hazards with time. To that end, early monitoring (see Section 21), for periods of at least 6 to 12 months, will be necessary, covering *inter alia*:

- Water quality and water levels in boreholes drilled for the site investigation, and in existing boreholes near the site, as well as in effluents and water courses, with, if possible, inflow/outflow water-balance studies
- Gas emission at selected sampling points (see Appendix 4)
- Settlement measured on a grid of levelling points
- Ground temperature.

15.4 Preliminary works

15.4.1 Off-site preliminaries

In order to minimise the exposure of personnel to site hazards the activities and the recording methods and forms should be planned beforehand. For example, plans for marking up site features, exploratory hole logs, and sample labels should be made ready off-site (see also BS5930 and DD175).

15.4.2 On-site preliminaries

Demolition and clearance. It is usual for this work either to have been completed before site investigation commences or to be included as the first phase of construction. However, if such work is necessary as a preliminary to site investigation, it should be carried out under the control of the engineer, and in such a way as to pose neither danger to personnel nor risk of spreading contamination and falsifying contaminant levels. An accurate record should be kept of the positions and nature of demolished structures before their foundations are covered over (Figure 21). Subject to the above, all demolition should be in accordance with BS6187: 1982, *Code of Practice on demolition.*

Figure 21 *Foundations exposed in groundworks (photograph by courtesy of Thorburn Associates)*

Wherever compatible with the proposed end use, inert or less heavily contaminated demolition rubble should be retained for use as bulk fill. The more hazardous materials may require disposal as 'special waste' to specially licensed tips, with the prior agreement of the appropriate waste-disposal authority.

Location of sampling points. Accurate setting out is essential, especially where contaminated ground may later have to be located for removal or other remedial treatment. The positions of trial pits and boreholes should be accurately located with reference to the site grid to facilitate re-location in the future. Temporary benchmarks should be established, from a permanent benchmark, in protected locations convenient to the sampling points.

Site facilities. Essential safety facilities, to the extent outlined as appropriate for site investigation in Section 23, should be provided on-site before the definitive field work starts. Other necessary facilities for site investigation include:

- Dry, cool store for samples and data
- Means of disposal of liquid and solid wastes
- Clean fill to top out pits, etc. when required.

15.5 Exploration methods of site investigation

15.5.1 Techniques required

The site investigation of derelict sites involves surface and sub-surface sampling from trial pits, trenches and boreholes, supplemented with chemical analysis or other laboratory tests.

An investigation of hydrogeological aspects will also be necessary. There are a number of specialised techniques which may usefully assist the sampling but which cannot be considered as substitutes for it. These include principally:

- Geophysical methods
- Aerial photography
- *in-situ* geotechnical tests
- *in-situ* chemical tests.

Early monitoring may be required for data on settlement and changes of groundwater level and flow, gas emission, ground temperature and combustion.

15.5.2 Trial pits and trenches

Trial pits and trenches are probably the most cost-effective technique for investigating derelict sites, but require great care to safety (Section 23.5.4). They are practicable for examination and sampling of materials in the ground to a depth of about 5 m, which in many cases will include the principal layers of contamination. They are also most useful in fills for examination of the lateral and vertical variations and for confirming the geometry of the boundary of the filled void.

Pits and trenches are convenient for locating and collecting samples and for studying the fabric of a fill and the form in which contaminants are present. The materials exposed in the pit or trench walls should be photographed in colour.

The arisings from a trial excavation should be set aside near the excavation, with care not to contaminate the trial pit or ground surface. They should be used for backfilling the excavation unless there is a risk of transferring deep-seated contamination to a level objectionably near to the surface. In such a case, clean fill is required for topping out. The excavation should be backfilled as soon as possible. Piezometers, gas- or water-sampling tubes, and probes (redox potential and temperature) may be placed in the pit or trench with the backfill laid around them.

One must be prepared to encounter large volumes of highly contaminated water entering, or even filling, the trial pits. This will necessitate pumping before backfilling and probably a supply of clean fill to top out pits.

15.5.3 Boreholes

Boreholes are required for exploration and sampling below about 5 m depth and for sounding the depth of deep fills. Boring avoids the necessity of large excavations and can provide both tube and bulk samples of the ground profile as in conventional investigation work. There is some inherent uncertainty in the exact depth of sample recovery and a lack of a visible profile of the material penetrated, unless continuous tube sampling is practised. Boring may lead to contamination of samples by materials from other horizons or by the fluids used for drilling. However, the alternative of deep pits is not usually economically justifiable.

As far as possible, water should not be used to assist boring. In some fills air-flush drilling has been successful. When the investigation is taken into rock beneath the fill, air-flush drilling should be used to prevent sample contamination or dilution from drilling fluids. With air-flush drilling there is a risk of blowing contaminated dust into adjacent areas. Well-maintained dust suppressors should be fitted.

In soft ground the borehole will need casing, to be extended as the bore proceeds. Where boreholes penetrate contaminated fill, the casing should be driven into the underlying formation to effect a seal before the hole is continued deeper. Smaller-diameter casing should then be drilled through a cement-bentonite plug into the formation (see Figure 22).

Whatever technique is used, precautions are needed to prevent contamination of the ground surface.

15.5.4 Hydrogeology

The groundwater regime effects the movement of pollutants within and from fill or contaminated land. The rise or fall of an established groundwater table may influence the

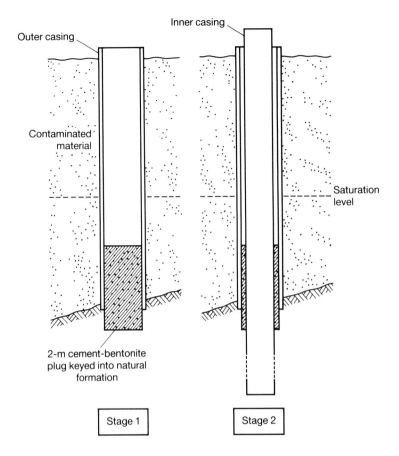

Figure 22 *Sealing boreholes through contaminated material*

settlement behaviour and the rate of decomposition of a fill, temperatures and decomposition products.

Early monitoring of groundwater in a fill should ideally continue for a period of at least twelve months. If that is not possible, the period should include the generally wetter months of February, March or April. Failing that, the recorded groundwater levels may not represent the worst conditions and evidence of any intermittent perched water tables may be absent. Observations should extend, as necessary, beyond the confines of the site.

A number of borehole installations may be needed to give a profile of the hydrostatic pressure with depth. For a more sophisticated analysis of the hydrogeology, 'downhole' surveys of flow in boreholes using flowmeters, combined with measurement of borehole diameter (caliper logging), fluid temperature and conductivity, can give indications of the levels at which there is contaminated seepage through the ground. Boreholes through contaminated ground for monitoring leachate/groundwater should be constructed so that only the leachate or groundwater from the zone under consideration can enter the borehole. They also should be sealed in such a way that polluted water cannot invade clean groundwater at other levels. Permanent lining tubes of metal are necessary, with grouted sealing plugs as shown in Figure 23. PVC tubes are suitable only where there is no danger of reaction between contaminant and PVC. The arrangement of the borehole will depend on whether it is to monitor leachate within a contaminated fill or the groundwater in the underlying materials. The configurations in Figure 23 are for the following purposes:

1. Sampling in fill only, followed by open-hole monitoring of fluids in fill
2. Sampling in fill and the unsaturated underlying strata, followed by open-hole monitoring of fluids in fill only
3. Sampling in fill and the unsaturated underlying strata. Suction probes may then be installed to sample gas or fluids in the unsaturated zone below the fill
4. Sampling in fill and the underlying strata including the saturated zone, followed by sampling the fluid in the saturated zone
5. Sampling in fill and the underlying strata including the saturated zone, followed by piezometer monitoring of fluids in fill and underlying saturated zone.

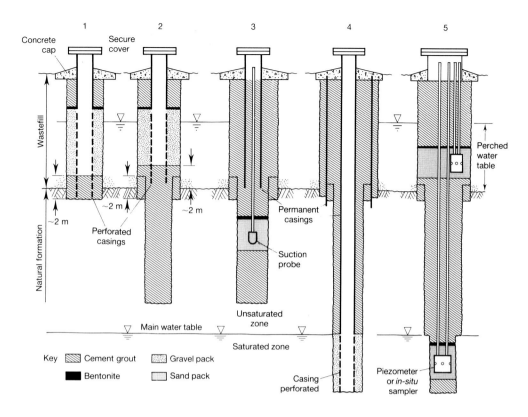

Figure 23 *Boreholes for groundwater monitoring in contaminated fill*

15.5.5 Geophysical methods

Geophysical methods provide indirect evidence of anomalies in the ground, caused by marked changes in physical characteristics. They can be a useful preliminary aid in the investigation of industrially derelict land and wastefill, in the approximate location of buried obstructions, and in defining the confines and consistency of fill. However, they are not a substitute for rigorous exploratory sampling in pits and boreholes.

With the aid of preliminary boreholes to provide geological control, geophysical methods may sometimes be usefully applied to the preliminary investigation of fill. The information gained may serve in designing a more efficient and economic layout of the definitive boreholes required for identifying the boundaries and depths of a fill area (e.g. the old highwall in reclaimed opencast mining sites) and, under suitable conditions, the depth to the water table. Another potential application is the detection of plumes of contaminated groundwater (leachate) from fill areas.

Geophysical methods require expert interpretation when applied to the heterogeneous mass of a fill, and can give valid results only when sufficient contrast exists between the materials on the two sides of an anomaly (for example, the boundary between infill and natural strata). Similarly, two parallel boundaries can be distinguished only if separated by sufficient distance. If, for instance, a perched water table in a fill is not sufficiently high above bedrock it may not be separable from the bedrock itself.

Of the two main techniques available, electrical resistivity methods may better serve to identify the depth to water table. Both seismic refraction and resistivity methods, on the other hand, are suitable for determining boundaries of steep-sided, infilled quarries if sufficient contrast exists.

Seismic methods are generally more accurate than resistivity methods, with errors in thickness of strata generally not exceeding 10–20%. With resistivity methods the accuracy is probably not better than ±20% on thickness. The main applications and methods, and their constraints, are presented in Table 13.

Table 13 Geophysical methods in surveying fill sites

Purpose	Geophysical method	Constraints
1. Measurements of depth to bedrock and steep boundaries	Seismic refraction	Fill too loose
		Highly contaminated fill, or fire or explosion risk may prevent use of explosives
		Non-explosive energy inputs may be weak against background 'noise' and require a 'signal stacking' technique
		Heterogeneous nature of fill renders interpretation of results difficult
		Weathered bedrock reduces contrast in seismic velocity
		Difficulty in drilling for explosive charges where these can be used
	Electrical resistivity	Heterogeneous nature of near-surface fills reduces the quality of interpretation of resistivity measurements
		Presence of a dry surface layer reduces conductivity in the vicinity of the probe (can be improved by soaking with saline solutions)
2. Measurement of depth to water table	Seismic refraction	As for (1) and when thickness of saturated zone less than 20–30% of fill thickness (three-layer problem with middle layer too thin)
	Electrical resistivity	Layering of fill or perched water tables can prevent analysis
		Water table near base of fill (middle layer too thin)
3. Detection of plume of contaminated seepage	Electrical resistivity	Width and thickness of contaminated layer should be not less than 20% of its seam depth
		Conductivity of the contaminated water should be an order of magnitude greater than the groundwater
4. Determination of the extent of a filled quarry	Seismic refraction	As for (1)
	Electrical resistivity	Heterogeneous nature of fill

15.5.6 Aerial photography and remote sensing

Aerial photography, particularly using infra-red film, can help detect polluted ground. Vegetation suffering from chemical contamination or methane gas does not reflect infra-red wavelengths as well as healthy vegetation. This deficiency produces a marked contrast on infra-red film, identifying areas for further investigation on the ground. Radio-controlled model aircraft may be used for aerial photography.

Expert advice is needed in interpreting the photographs and all conclusions should be checked by ground study. There are difficulties in interpretation, e.g. different plant species are affected differently by methane. Other contaminants or soil deficiencies can cause distress very like that from methane.

Ideally, photographs should be of the vertical type, to facilitate the transfer of information to maps or plans. False-colour images from satellites such as 'Landsat' are not usually of sufficient resolution. Infra-red thermal-imaging from the air or a convenient high vantage point will detect surface temperature differences which may indicate underground combustion or other exothermic reactions.

Recently multi-spectral imagery has been applied to the investigation of contaminated sites. Coulson and Bridges (1985) present two interesting pilot studies.

15.5.7 In-situ *geotechnical tests*

On derelict sites, other than fill, geotechnical properties may not have a decisive influence on the future use of the site. On fill, however, the geotechnical nature and state of the fill materials themselves are usually determining factors, because they control potential settlement.

The common *in-situ* geotechnical tests described in CIRIA Special Publication 25 (Weltman and Head, 1983) (standard penetration test (SPT); cone and dynamic penetrometer; pressuremeter; *in-situ* density; plate bearing; and permeability) are of limited application to fill sites, due to the heterogeneous nature of fill and the time-dependency of self-weight settlement. Some of the tests may be used in opencast waste and granular fills, provided that the fraction of large cobble or boulder-sized material is small. In general, wastefills and landfills are not amenable to these tests.

Some older domestic refuse (pre-1950) may be sufficiently ashy and compacted to give valid results from penetration and bearing tests. The condition of the fill should be ascertained from trial pits or boreholes before testing is carried out. *In-situ* testing (e.g. SPT) can then give useful comparative information regarding the state of the fill.

For shallow fills (not more than 3–4 m of fill below the founding level), large-scale plate loading tests can indicate the compressibility of the fill and its suitability for footings. The skip-loading test described by Charles and Driscoll (1981) is particularly appropriate where light industrial or domestic buildings are proposed on shallow fills not containing domestic refuse. A summary of suitable *in-situ* tests is given in Table 14.

Table 14 Summary of *in-situ* geotechnical tests for fill

Test	Suitable fills	Precautions and notes
Bulk density	Most types except recent domestic or industrial refuse	Large samples are necessary in many fills, especially if there are wide size ranges and large maximum particle sizes. A sand replacement method cannot generally be used in loose, uncompacted fills, or fills with coarse granular particles. In fine-grained cohesive fills the core cutter method may be used
Permeability	Essentially only non-cohesive natural fills below the water table	Test results may be erratic, as a result of uneven compaction, heterogenity of the fill material and wide variation in particle size. The tests can be used as indicators of variation in such properties if calibrated against other properties (for example, dry density)
Penetrometer (SPT, CPT, DPT)	Natural homogeneous fills	Suitable in gravel sized or finer
Pressuremeter	Natural fills and some landfill refuse of an ashy type	Problems may occur in granular fills with large particles (>75 mm) which can prevent the insertion of the driven probe casing. Very dense materials are also unsuitable. In wastefill, the results may be of comparative value only
Plate bearing: up to 1 m diameter	Relatively homogeneous fills with grain size less than 75 mm	The tests should be carried out over a range of depths corresponding to the stressed zone. The results are of limited value in typical heterogeneous fills
Area loading: long-term	Most fills, provided there are no large voids or extremely large particles	This test is typified by the 'skip loading test' and is suitable for many fills—for which valid settlement data are unobtainable by any other means. However, this only gives design information in shallow fills (less than 4 m deep)

15.6 Sampling strategy for contamination

This section deals with the basic principles underlying sampling on contaminated sites. The chemist and the engineer should work closely together in determining the sampling strategy. Specialised techniques for the collection and handling of samples are described separately in some detail in Section 16.

15.6.1 Samples required

Surface sampling of solid materials should always be undertaken. Where the initial assessment leads the engineer to suspect sub-surface contamination, sampling should be carried out to a depth at which any contaminants will no longer affect the proposed end use, but taking account of the potential for groundwater pollution. On fill sites, the full depth of the fill should be sampled. Liquids and gases should be sampled on sites where hazards from gas emission or leachate are suspected.

15.6.2 Sampling strategy

Few derelict sites are uniformly contaminated all over. Contamination may be variable across the site, or present in pockets or otherwise non-uniform. Fairly substantial areas may be free of significant contamination. The sampling strategy should take account of this possibility of variation and should also not lose sight of the desired end use.

Sampling strategy for general geotechnical purposes is well covered in BS5930 and Weltman and Head (1983). For contaminated sites, the provisions of DD175 *Code of Practice for the identification of potentially contaminated land and its investigation* (BSI, 1988) should be carefully followed.

Sampling intensity should be specified to satisfy in the most economic manner the following premise:

> All contamination which is of sufficient intensity and magnitude to pose a significant threat to a chosen end use will have to be treated, sooner or later, or the project abandoned.

Ideally, the sampling should be rigorous enough to ensure that every significant presence of contamination is detected. Contamination missed during sampling may come to light later during the building works. It will almost certainly then prove considerably more costly to treat, may cause expensive delays in site work, and may enforce a change of plan or even abandonment of the project.

In practical terms, however, site investigation is limited in what it can achieve, unless the number of trial pits, boreholes, etc. is to be prohibitively expensive. Moreover, the more extensive earth-moving works of the subsequent reclamation stage have inherently more opportunity of bringing to light previously undetected hazards (see Section 5.3).

The economic balance between these conflicting factors leads to the current practice of locating sampling points on a systematic square or rectangular grid (orthogonal or staggered). At all places the resolution of the grid should be such that the area included between sampling points is no greater than the largest area of contaminant that could be dealt with economically if left undiscovered until a later stage (Beckett and Simms, 1985).

The precision of an estimate of the contaminated state of a site depends on the fineness of resolution of the sampling grid. Increasing fineness over the whole site, however, yields diminishing returns in terms of overall project economics.

A sampling plan has to be flexible to suit the given site and proposed end use(s). Rigid rules cannot apply and guidance only can be given. Two important factors influence the degree to which the sampling plan can be defined:

1. Information obtained from the initial assessment (site history, nature and location of foreseeable contamination)
2. Analyses required (depending on end use, see Table 15, for example).

The more complete the information obtained from the initial assessment, the more effectively can sample points be located. Where the initial assessment has indicated the likelihood of areas of greater or lesser contamination, the sampling grid can be adjusted for precision and economy, in particular by greater frequency of sampling in areas of suspected high contamination or to prove that an area is 'cleaner'. However, the reverse procedure, reducing the intensity of sampling on the supposition of non-contamination, should only be permitted with the greatest caution. It is essential not to prejudge the state of contamination. Even where preliminary evidence may point to freedom from contamination, migration of contaminants should always be suspected, in the absence of hydrogeological data.

Individual samples should be taken at each point and at each defined level. Samples should not be mixed to give average values of contamination.

15.6.3 Number of sampling points

The detailed investigation and appraisal of a derelict site are interlinked and iterative procedures, in which feedback of information is used to pinpoint aspects or areas requiring further investigation. Normally, preliminary sampling is carried out on a regular grid generally at 25- to 50-m spacing. The subsequent examination of the samples serves to indicate areas where further, more intensive sampling may be desirable.

The initial assessment may already have given an indication of areas where contamination is to be suspected from the past history of the site. In this case the preliminary grid may have to be intensified around areas of suspected contamination.

Generally, however, where no prior information is available for, say, an uncontrolled waste filling, the preliminary sampling should be on a uniform grid covering at least the whole site. If the extent of the fill is not known accurately, the grid should extend into known original ground. Sampling for methane should be extended beyond the confines of a wastefill to a distance which reflects the scale of the gas source and the permeability of the surrounding ground mass.

The density of sampling in the preliminary grid may be expected to vary inversely with the size of the site. Recommended minimum numbers of sampling points given in DD175 are:

Size of site (ha)	Minimum number of sampling points
0.5	15
1.0	25
5.0	85

A grid spacing of more than 50 m will be of little value except for very large, unobstructed fill sites.

Where defined areas of contamination are suspected from the site history of an old industry, the preliminary sampling density can be adjusted to provide sub-grids of more closely spaced points. Generally, however, sampling at a finer grid resolution will take place only after the results of the preliminary sampling have been appraised. Spacing closer than 10 m will normally be neither necessary nor economically justifiable. Recent studies (Smith and Ellis, 1985) have shown the optimum for a gas works investigation to be around 25 m.

Fine-resolution sampling may be unjustified in certain cases including:

- Large volumes of material so obviously and uniformly contaminated that the same drastic treatment will be required throughout the whole mass
- An area so obviously badly contaminated that the remedial decisions do not require quantitative data
- Small, apparently uncontaminated pockets, within irretrievably contaminated areas, and too small to be profitably developed
- Parts of a site so badly obstructed by massive foundations, etc. as to be obviously unattractive for building works. (Sampling might be devoted instead to providing the boundaries of the acceptable areas.)

However, before opting for restricted sampling on the ground of economy the engineer should consider carefully the consequences of a future reversion of the site to a more sensitive end use.

Fine-resolution sampling may be necessary in cases where contamination is so serious that the contaminated ground must be removed to a tip licensed to receive it. Accurate data will be necessary to establish the quantity to be removed and accepted by the tip operator.

On fill and similar sites, where sampling for methane may be considered necessary, a preliminary 'spiking' survey of the sub-surface (500 mm depth) on a grid spacing between 10 m and 25 m may provide a useful guide for planning the deeper sampling provided that the spikes fully penetrate the cover. The deep probes for measuring and monitoring gas emission should be located on the overall grid in areas where the development is likely to be concentrated and/or where the 'spiking' tests have detected methane. Boreholes should normally be spaced at 25–50 m. Carpenter et al. (1985) Pecksen (1985) and, in particular, Crowhurst (1988) have dealt in more detail with the design and execution of a programme or methane testing (see also Appendix 4).

15.6.4 Depth of sampling

Site sampling, preferably in trial pits, is required to define the horizontal and vertical distribution of contaminants to such depth as the material may pose a hazard to targets exposed by the proposed end use, including the environment, groundwater, etc. A useful guide for the preliminary sampling, where the depth and nature of the contaminant may be unknown, is to dig the trial pits to a depth of at least 3 m and take 'spot' samples at 0.15 m, 0.5 m, 1.0 m, 2.0 m and 3.0 m (Carpenter *et al.*, 1985) with additional samples to reflect changes in strata as visually different materials. For more detailed sampling in sub-grids these depths may still be suitable or they may be modified, as discussed below.

Migrating contaminants. Sampling should be taken down to such depth that upward migration of contaminant or aggressive chemical will not present a hazard to a sensitive target (nominally 3–4 m).

Combustible material. Where combustible material is suspected, sampling should prove the full depth of such material.

Putrescible materials. Landfill gas can be produced and migrate from any depth. Sampling should prove the full depth of such tipped materials down to 'clean' ground.

Deep sampling from boreholes may be required to establish:

1. The geometry and extent of a fill site, and the nature of the fill material
2. Distribution of gas in the ground
3. Extent of combustible material
4. Engineering quality of underlying ground, especially for medium depth (5–10 m) fill lying on weak materials
5. Hydrogeological data, including evidence of groundwater contamination and leachate
6. Depth, extent and condition of oils, solvents, tars or other organic liquids spilt into permeable ground.

For items 1–3 the bore should be taken to a depth of 1 m into the original ground below the fill. For item 4 the bore should continue to a depth determined by conventional geotechnical considerations.

Where contaminated groundwater is a problem on a permeable sub-stratum the water authority may require boreholes to prove the bottom of the lowest permeable stratum to which the leachate may have penetrated (item 5).

15.6.5 Sampling techniques

The techniques for taking disturbed samples for chemical analysis differ from those for normal geotechnical purposes, and the engineer should take advice from the chemist in the team regarding the specification of procedures.

Techniques for taking, handling and transport of samples depend on the nature of the material and the investigation method being used. Special requirements for sampling on a derelict site are outlined in Section 16.

15.7 Chemical analyses

15.7.1 Basic strategy

The engineer should rely on the advice of the chemist in the team for all aspects of chemical testing, but should, however, ensure that the analytical strategy conforms to the basic concepts set out below, especially on limiting the work to the essential minimum.

The quality control and quality assurance regimes proposed by a tendering analytical contractor should be checked for adequacy. This is a useful guide to the calibre of an analytical contractor.

It is important not to prejudge an unknown site, even though testing for the full range of possible contaminants is rarely necessary. Considerable judgement is required to avoid execution of superfluous or inappropriate analyses and, equally seriously, inadequate coverage of an unknown site. The basic strategy must be continually reviewed to ensure cost-effectiveness. ICRCL Note 59/83 (2nd edn 1987) is a valuable aid to the engineer in this regard.

15.7.2 Specification

Chemical reactions taking place within the ground are often slow, complex, and influenced by many variables, so that their evaluation and prediction are fraught with uncertainties. Universally agreed and accepted standards do not exist for the laboratory analysis of samples of ground contaminants or their correlation with conditions on-site.

The role of the chemist in the professional team is most important: the chemist should specify and control the laboratory work. It is not good enough simply to ask a laboratory to execute an analysis for a certain contaminant without qualification as to how it is to be carried out in detail. Without specific directions, an inappropriate sample preparation and/ or analytical method may be employed, and the result may be open to misinterpretation or incorrectly compared with results from another laboratory.

It is unrealistic to expect the analyst to interpret the results in isolation from the other data on the site investigation. Under such circumstances, the analyst can give only a restricted comment on the sample examined with no attempt at generalisation as to the site as a whole. Accordingly, the specification for analysis drawn up by the chemist has to be precise and detailed.

15.7.3 Scheduling

The chemist should schedule the various samples for analysis and specify the contaminants for which analysis is required. This should be done, on the basis of the strategy outlined, in stages gradually confirming or eliminating contamination in a logical sequence.

A large range of chemicals may exist in the ground on a contaminated site but only a few are likely to be relevant or harmful to a given end use. In scrapyards, for instance, high metal concentrations are unlikely to affect industrial development (except for possible bi-metallic corrosion effects) but could be serious for horticulture or children. On the other hand, mineral oils and PCBs, also present in scrapyards, could be hazardous for building development.

The interplay (site/hazard/end use) illustrated in Table 2 (inside back cover) is fundamental to all aspects of site appraisal and, in particular, to decisions on what chemical analyses may be necessary in any given situation.

As explained below, first-stage analysis should be restricted to those substances known to be harmful to the chosen end use and whose presence may be suspected from the site history. The programme of analysis should be modified from time to time if the contaminants brought to light suggest changes in end use, remedial measures, etc.

15.7.4 Analyses related to end use

Table 15 illustrates in greater detail a part of the relationship shown in Table 2 (inside back cover) regarding the sensitivity of various targets to commonly occurring contaminants. It suggests a priority sequence of contaminants for inclusion in a programme of analyses, with modifications, if necessary, to suit the known history of the site and its desired end use.

Two degrees of hazard are assumed:

1. *Listed contaminant hazards* (from Table 3 of DD175: 1988 and from Table 2 of ICRCL Note 59/83, 2nd edn. 1987). A programme of analyses for a given end use should include such of these hazards as are shown in Table 15 to affect the relevant target(s) exposed by the end use
2. *Non-listed contaminant hazards* which may be locally significant but are generally less likely to pose a problem for a given end use. Where evidence points to the possible incidence of such hazards it may be necessary to include them in the analyses appropriate to the end use.

Table 15 Contaminant hazards related to end use

	Targets and associated hazards[1]						
	1 **Site works**	**2** **Water resources**	**3** **Buildings and services**	**4** **Buildings and services**	**5** **Garden amenities**	**6** **Allotments, agriculture**	**7** **Plant growing**
Critical Contamination[2]	**Harm to workers**[3]	**Leachate pollution**[4]	**Fire and explosion**	**Corrosion or expansion damage**[4]	**Harm to children**[5]	**Crop uptake of contaminant**[4]	**Phyto-toxicity**[4]
Arsenic	○	○			●	○	
Cadmium		○			●	●[15]	
Lead		○		○[11]	●	●[15]	
Cyanide	○	●			●[14]		
PAH[6]	●		○		●		○
Phenols	●	●		●[11]	●		○
Copper		○		○[11]		○	●[16]
Nickel		○		○[11]			●[16]
Zinc		○		○[11]			●[16]
Sulphate		●		●[12]	●		○
Sulphide		○		●			
Chloride		○		●			○
Mineral oils	●[8]	○	○	●	○		
Ammonium ion				●			
Methane	○[9]		●[9]				○
Sulphur			●	○[13]			
Combustibles	○		●[7]				
Oily, tarry substances	●	○	○	●			○
Asbestos	●	○[10]					
Radioactive materials	●	○			○	○	
Soluble metals		●		○	○		
pH	○	●		●		●	●
Solvents and organics	○	○	○				
Boron		○					○
Expansives				○[17]			

● Listed contamination hazards (from BS.DD 175: 1988, and ICRCL Note 59/83 2nd edition 1987).
○ Non-listed contamination hazards (see Section 15.7.4).

Notes:
1. Combinations of hazards may need investigation.
2. Local factors may involve other contaminants.
3. Workers at risk from contact, inhalation, explosion, fire and ground pH.
4. Soil pH affects incidence of these hazards.
5. Children at risk from inhalation, ingestion and contact.
6. PAH: polynuclear aromatic hydrocarbons.
7. Combustibles include: coal, oils, tars, refuse, rubber, etc. Determine: calorific value, loss on ignition, ash and volatile contents, total organic carbon.
8. Mineral oils are industrial hygiene problem, with or without PCB contamination.
9. Continuous on-site monitoring.
10. Hazard if run-off carries friable asbestos into streams.
11. Electrolytic action if free metal present.
12. Total and soluble sulphate. (See also Section 11.4.4).
13. Corrosion induced by sulphur-consuming bacteria.
14. 'Free' cyanide.
15. Total and 'plant-available' metals.
16. Total, 'plant-available' and 'zinc equivalent' values.
17. Expansives include: blast-furnace and steel slags, mining spoils and pyritic materials. Determine: total sulphur and acid-soluble sulphate, calcite, and calcium as Cao.

Consideration should also be given to the inclusion of other contaminants if the site history has identified former uses likely to have introduced them (refer to Tables A1 or A2). Thus Appendix 3 sets out further chemical contaminants which may require consideration on specific types of site. Moreover, ICRCL Note 59/83 gives 'trigger values' (see Section 17.4) for additional substances, including chromium, mercury, selenium, boron, complex cyanides and thiocyanates. However, as a basic principle, analyses should be limited in the first instance to the listed contaminant hazards as above. The chemist's advice should always be sought before any attempt is made to undertake analyses beyond the listed items.

A given end use will usually expose more than one type of target to contaminant hazard, so that several columns of Table 15 will be of significance. In all developments, irrespective of end use, site workers and water resources (Columns 1 and 2) must be considered as targets of prime importance. The relative significance of the remaining columns will depend on the type of development. Thus, on a purely building project (industrial or housing) without gardens or similar amenities Columns 3 and 4 will have over-riding importance, possibly to the exclusion of Columns 5 to 7. In contrast, in a development involving gardens, amenities or plant-growing facilities, Columns 5 to 7 may be of equal or greater importance.

These considerations also apply separately to small parcels of land, amenity areas, domestic gardens or similar micro-areas, which may be included within the confines of an otherwise predominantly building development. In a global building project various areas may require differing priorities of analyses to suit specific targets.

The application of Table 15 in determining a programme of analyses requires an appreciation of the factors discussed below.

Health and safety of site workers (Column 1). This must always be of paramount importance, from the first entry on the site for reconnaissance through to completion of construction. Workers can be at risk from skin contact, inhalation, and fire and explosion hazard. Moreoever, when the ground pH is below about 2 or above 11, conditions can be unpleasant and harmful. But sampling and analysis cannot be expected to *detect* all hazards in time to ensure complete absence of risk to investigation personnel. Safety measures must be available for dealing with unexpected hazards during the early stages.

Hazards to water resources (Column 2). There is growing concern to safeguard natural water resources and aquifers. The statutory authorities may forbid developments unless the site investigation has been sufficiently rigorous to ensure the avoidance of this hazard.

Fire and explosion hazard (Column 3). If either of these hazards is found to be present it may rule out the site for any type of building work, so that other tests become irrelevant.

Corrosion of building materials (Column 4). If foundations, buildings or services are installed in disregard of potential aggressive chemical attack, subsequent repair work may prove impracticable. Corrosion of metallic items may also result from electrolytic action due to buried metal contamination. When in the form of discrete components, free metals may perhaps be recognisable without analysis. On the other hand, where such contamination is present as metallic dust, analysis may be essential.

Hazards affecting health and plant growth (Columns 5, 6 and 7). These will be important if the development includes gardens and similar amenities or plant-growing facilities. They may be of less significance in developments involving only buildings and hard surfacing, but even these areas will eventually require excavation works for maintenance of services, etc., with consequent hazard to workers. Furthermore, even purely building projects may have open, non-covered spaces. Poor vegetation cover in such areas can increase the risk of inhalation. In areas of exposed ground children can be at risk from inhalation and ingestion of contaminant, as well as through skin contact, whereas adults (gardeners, maintenance workers, etc.) are more likely to suffer through skin contact and inhalation.

15.7.5 Scheme of analyses

Figure 24 illustrates a scheme of analyses for solid and liquid samples from contaminated sites. It is derived from the Final Report of the Policy Review Committee of the DoE Co-operative Programme of Research on the Behaviour of Hazardous Wastes in Landfill Sites (DoE, 1978).

15.8 Groundwater Investigation

It is normally the responsibility of the developer to carry out a full investigation and appraisal of the leachate hazard on a site proposed for development. The developer is obliged to satisfy the water authority (and perhaps the waste-disposal authority) as to the ultimate safety of the local water resources and effluent discharges from the site.

Pollution and leachate investigations have been extensively dealt with in the WRc Technical Report TR91 (Naylor *et al.*, 1978) which should be referred to in all cases where leachate is a cause for concern. Such investigations start by determining the extent and type of pollution associated with the site and relating these data to the history of tipping on the

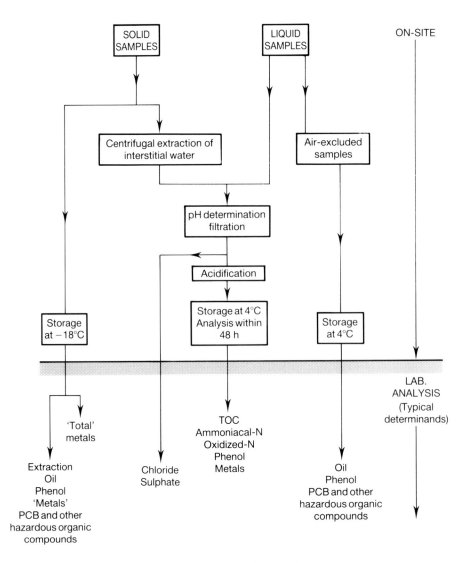

Figure 24 *Scheme for analysis of solid and liquid samples*

site, modes of operation, types of waste, local hydrogeological and climatic conditions, so as to permit predictions of future concentrations and movements of the contaminants.

The planning, detailing and execution of a campaign of leachate investigation, as well as the appraisal of the results, are normally the province of a qualified hydrogeological or similar consultant, specialising in contaminated fill sites. Investigation requires the monitoring of borehole water samples and water levels, flow rates, rainfall and effluent outflow. Ideally, monitoring should continue for at least 12 months. If that is not possible, recordings should be taken at least when the groundwater levels are at their highest. In the UK, this is usually during March or April. The results are used to establish a water balance and data on leachate enrichment during percolation through the fill or contaminant sufficient to define the volume, flow pattern, and composition of the leachate to be expected.

15.9 Combustibility investigation

15.9.1 Sub-surface temperature surveys

Even without visible effects on the surface (smouldering, charring, smoking), hidden combustion may be proceeding underground and may be betrayed by locally elevated temperatures in the ground. Where thermal activity is suspected, temperature surveys should be carried out by instruments located on a grid, initially 25–100 m square and later at increased frequency at suspected or discovered hot spots.

The sub-surface temperature distribution may be directly observed on instruments lowered into boreholes or by probes driven into the ground. The instruments should be left at the appropriate depth for about an hour before reading. In boreholes, air or water movements may falsify the readings.

Instruments and probes should be based on thermocouples with remote recording and signal telemetry facilities. Instrument probes may be left in position for future long-term monitoring of the sub-surface temperature excursions.

15.9.2 Surface temperature surveys

Thermal-imaging infra-red equipment, either hand-held or airborne, will provide a rapid scan of surface temperatures. Under favourable conditions, airborne equipment can distinguish temperature differentials as little as 1°C and can thus indicate sub-surface combustion at considerable depths.

Thermal-imaging equipment is complicated and expensive, and demands considerable skill in use. The sensor requires cooling by argon gas expansion or by liquid nitrogen, depending on size.

15.10 Investigation of slags, mining spoils and building wastes

Section 11.4 explains how chemical instability can render these materials troublesome on two counts:

1. Expansive reactions
2. Sulphate activity (actual or potential).

The main objectives of investigation will be to determine the geometry of the fill (i.e. depth and lateral extent) and to assess its content. In general, these materials are not easy to bore through, particularly building wastes and metallurgical slags. Recourse to rotary percussive drilling may be required. Identification of the materials in hand specimens is necessary together with respresentative sampling for chemical analysis (see Table 15). Investigation of the commoner fill materials discussed under this heading should take account of the following points.

Building rubble. Components containing gypsum may be recognisable by inspection, otherwise sampling and analysis for sulphate will be required. Rubble arising from demolition of chimneys will almost certainly be contaminated with sulphate.

Iron and steel slags. These are difficult to distinguish. Major oxide analysis can usually give positive identification, but X-ray analysis or microscopy may be needed. Once identified, no further testing of a steel-making slag is required. In the case of iron-making slags, determinations to BS1047 of sulphate and total sulphur contents should be made.

There is no standardised method for the assessment of the expansive potential of these materials. BRE can provide advice, however, on accelerated testing based on storage at elevated temperatures.

Pyritic shales. Conventional sulphate testing to BS1377 will not be sufficient to reveal the potential of these materials for troublesome oxidation, with possible expansion hazard and sulphate/acid production. For proper appraisal, recognition of pyrite in the fill material is necessary. If known, it is also advantageous to inspect the source of the material, especially if it is one of the natural formations listed by Taylor and Cripps (1984) (see Section 11.4.3). The detection of jarosite or gypsum in laminations should be a warning of potential expansivity problems.

Samples from the investigation should be analysed for sulphate and total sulphur, calcite, and the tendency of the material to oxidise on exposure under acid conditions favourable to pyrite-oxidising bacteria.

15.11 Investigation of landfill gas

For the interpretation of a landfill gas presence, information will be required, sufficient for the production of a site map showing the variations (over at least a six-month period) in the extent of the areas affected, the concentration and emission rates of the gas. Appendix 4 outlines the necessary procedures.

15.12 Reporting

The engineer should prepare (or have prepared by the investigation contractor) a factual report on the site investigation. This should be a true and complete record of all work carried out, and should include:

- Plans showing location of all trial pits, trenches, boreholes and sampling points
- Plans showing positions of all visible surface contamination and old foundations
- Details of all field testing and monitoring carried out
- Geotechnical logs of boreholes and pits with the added definition of colour and/or other signs of chemical contamination
- Details and results of all geotechnical testing undertaken
- Details and results of all chemical analyses carried out.

The report should state precisely the procedures adopted and the conditions under which the analyses were conducted. The detail should be sufficient to permit realistic interpretation of the results by other chemists not privy to the testing.

16. Sampling techniques

This section supplements the basic information given in Section 15.6. It sets out what is considered to be good practice in the collecting, handling, preservation and transport of contaminant samples from a derelict site. Other sampling techniques and variants may be acceptable alternatives, provided they comply with the basic principles set out below.

Specialist advice may be needed for particular contaminants, e.g. volatile organics could be entirely lost from a soil sample stored in a polythene bag.

The following notes should be read in conjunction with DD175 which, being the current statement of good practice, should be followed. During all sampling operations, whether from trial pits or elsewhere on the site, the relevant safety precautions stated in DD175 and in this publication have to be rigorously observed.

16.1 Sampling of solids

16.1.1 Equipment and handling

Hand sampling tools should preferably be of stainless steel and should be cleaned each time before taking a sample. Soil samples should be placed in wide-mouth, sealable, watertight containers of not less than one-litre capacity. The container should be of a material not liable to react with the contents. The body of the container, not the lid, should be indelibly marked in accordance with a consistent system for identification.

16.1.2 Surface samples

These are required for an appraisal of the variation in surface condition over the site and are normally taken from the top 150 mm of the ground. They should be taken before any levelling or site clearance has seriously disturbed the ground at the sampling point. The surface should be sampled at the grid intersections or other points preselected by the sampling strategy. Samples should also be taken representative of any unusual materials/ conditions on the site. In that case, the sampling point must be accurately surveyed and located on the site plans.

Before any surface is considered representative, due allowance should be made for weathering or other disturbance in the top 150 mm. The depth of the layer sampled should be consistent over the site. Surface samples may conveniently be taken with hand tools. Surface plant roots should be loosened with a fork to leave the soil *in-situ*, and the sample removed with a trowel.

16.1.3 Sampling from trial pits

The suggested methods of excavating trial pits and taking samples depend on the required depth of the pit:

Depth	Method of excavation and sampling
1 m	hand excavation by pick and shovel, with sampling by trowel from sides of pit. Only economic on small sites with less than, say, 10 pits
3 m	wheel-mounted back-hoe, with hand sampling by trowel from parent material in excavator bucket or, with long-handled tools, from the pit itself
5 m	track-mounted excavator, with sampling as for 3 m depth.

As a general rule, in pits more than 1 m deep, samples should be taken from the excavator bucket without entry of personnel into the pit. If entry is essential, the precautions of Section 23.5.4 should be followed. In such cases, sample tubes may be driven into the pit walls or, in dense fills, small samples may be obtained by use of a hand-driven auger.

Samples should be taken and placed in sealed containers as soon as possible after excavation, to minimise possible chemical or physical changes through exposure to air.

16.1.4 Sampling from boreholes

All material should be described as soon as it is acquired from the borehole and samples placed immediately into sealed containers. Further samples may be washed or otherwise prepared to aid the identification of the components. Alternatively, material may be left in the sampling tube, the ends of the tube filled with wax, or screw-capped, and the whole submitted to the laboratory for examination and analysis of the contents.

Should it be necessary to use water or other lubricant in the drilling or boring process this should be clearly stated on all relevant records and on the sample containers.

On completion of sampling the boreholes should be backfilled and sealed with a substantial clay plug or grouted with a cement and bentonite grout from the base upwards.

16.1.5 Contamination of samples

The area round a borehole or pit must be kept well drained and free of extraneous contamination from other sources. Borehole casings should project above ground level sufficiently to prevent surface water flowing into the hole. The casing must be advanced with the bore to minimise migration of leachates or interstitial water to lower levels.

Where samples or cores are taken from a stratum underlying a wastefill to check possible contamination of aquifers, the cores may become contaminated with borehole water or lubricant from the drilling equipment. If this occurs, the outer 20–30 mm of core should be discarded (Naylor *et al.*, 1978). To allow for this, cores should be not less than 100 mm in diameter.

16.2 Sampling of liquids

16.2.1 Equipment

Scoops, wide-mouth jars and vacuum pipettes are suitable for surface water sampling. Borehole sampling requires balers and mechanical samplers. Special samplers are available for sampling from standpipes, which is the best method to avoid contamination from on-site sources. Figure 25 shows borehole samplers. Automatic liquid samplers with pre-evacuated containers are available to take incremental samples at selected time intervals or to give average samples over selected periods.

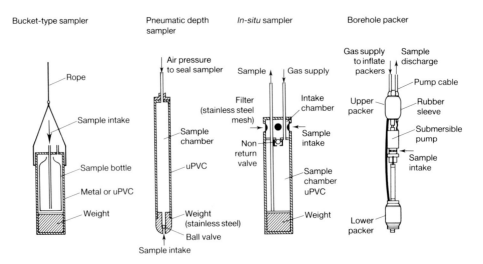

Figure 25 *Groundwater samplers*

Liquid samples should be filtered on-site and contained in glass or plastic (polypropylene or polyethylene) jars of at least one-litre capacity and securely labelled. Plastic materials should be selected with care if there is danger of selective absorption (of organic liquids, for example). Containers and lids should be rinsed three times with the sampled liquid before filling.

16.2.2 Sampling surface liquids

Liquid on, or accessible from, the surface may be collected in a scoop or by vacuum pipette. Automatic samplers can give additional information from flowing fluids. From a mixture of liquids, such as oil and water, several samples should be taken where single representative samples cannot be obtained (Figure 26).

Figure 26 *Seeps of oil (photograph by courtesy of GLC)*

16.2.3 Sampling liquid in trial pits

Liquid entering during the excavation of a trial pit should be sampled at the time and may provide reliable evidence of the state of ground contamination. It is not necessary to obtain 'clean' samples from a pit. The depth of the liquid entry should be recorded as well as colour and odour. Short standpipes may be installed in trial pits for the collection of representative samples of groundwater.

16.2.4 Sampling groundwater in boreholes

Water and other liquids standing in or seeping into a borehole may be recovered by a baler or sampled *in situ* by depth-controlled equipment. In the case of a high fluid yield, a pump may be installed and the delivered fluid sampled. Pumping tests, carried out according to BS5930, may give valuable information on ground permeability and possible migration of contaminants. Flow or recovery rates are important in assessing the validity of liquid samples collected from boreholes. Wells should be purged prior to sampling.

Liquid samples may be recovered from cores (obtained by rotary drilling or cable percussive techniques) by centrifuging or compressing in a specialised laboratory.

16.2.5. Handling and testing of liquid samples

Most liquid samples, once collected, will start to deteriorate, particularly by oxidation, and should therefore be analysed with the minimum of delay. Deterioration may be retarded by chemical treatment, storage in pressure vessels, or by conserving at about 4°C. Known hazardous samples should be so labelled and precautions for transport arranged with the relevant statutory authority.

An on-site laboratory has advantages in permitting rapid analysis of unstable liquids before major changes can occur. Moreover, data become immediately available for further direction of the field sampling. Site laboratory tests should be confirmed by off-site tests in a fully equipped laboratory.

Certain properties of water (including temperature, pH, dissolved oxygen content and redox potential) are quickly affected by contact with the atmosphere. The properties cited can be measured *in situ* by electrical or electro-chemical sensors. Sensors should normally be sealed into the ground at appropriate depths in boreholes.

16.3 Sampling of gases

16.3.1 General

Gases may be sampled in the free atmosphere, but the results are of little value. Somewhat more useful preliminary data may be obtained by adventitious sampling in trial pits, drains, rabbit holes and other confined spaces. These are no substitute for data obtained from a preliminary spiking survey followed by long-term monitoring from sampling tubes sealed into backfilled boreholes or inserted as probes into voids in the ground. The presence of methane in the ground can sometimes be inferred from its effect on vegetation as revealed by infra-red photography (see Section 15.5.6).

16.3.2 Handling of gas samples

A few gases, including hydrogen sulphide, may oxidise or otherwise react if taken to an off-site laboratory and are better analysed *in situ* or in a site laboratory. For most other gases found on contaminated or landfill sites, transit times to an off-site laboratory are not critical, provided the gases are stored in non-reactive containers.

Glass containers are suitable but require protection against breakage; rubber or plastics are not recommended. Metal containers can hold gas under pressure. Handpumps are acceptable for delivering gas samples to containers. Sampling tubes in boreholes and voids should ideally be of stainless steel; plastics may be acceptable for some gases. The advice of a chemist should be followed in all decisions on specific gases to be sampled and on materials required for equipment and containers.

16.3.3 Sampling in free atmosphere

This is of very little value in assessing the magnitude of the gas hazard. For the day-to-day safety of site personnel, toxic or flammable gas in the atmosphere, in open excavations, trial pits or closed spaces may be detected with portable instruments (hot-wire katharometers, gas-detector tubes, infra-red analysers).

If samples are required for laboratory analysis, a record should be kept of wind speed and direction, as well as barometric changes over the previous 24 hours (gas emission from the ground can increase on a falling barometer).

16.4 Sampling biological matter

16.4.1 Micro-organisms

Sampling equipment may include spatulas, pipettes, scoops and swabs and large numbers (say, 100/ha) of 10-ml glass or plastic containers. Samples requiring the preservation of particular conditions (for example, BOD, anaerobic, etc.) may demand specialised containers. All equipment and containers must be pre-sterilised and sealed in sterile packages until use.

Samples may be required from surface solids or liquids and from materials taken from trial pits or boreholes. Sampling demands specialised microbiological techniques by skilled personnel to obtain uncontaminated and representative samples.

Samples should be protected from sunlight and transported to a laboratory within a few hours. Containers holding pathogenic material should be clearly so labelled and should be sealed in a secondary leakproof plastic bag. Transport time may be extended to 30 hours provided the samples are maintained at not more than 4°C.

16.4.2 Macroscopic flora and fauna

Sampling equipment may include spades, forks, secateurs and metal counting rings, preferably of stainless steel, carefully cleaned each time before use. Containers may conveniently be polyethylene plastic bags of 1–100 litres capacity which should be clean and unused and sealed until required. Foliage samples should be protected from dust, fumes or liquids, and should not be washed on site; airborne surface deposits are often an important source of contamination on foliage.

Transport conditions are not normally critical unless living macroflora specimens are required.

16.5 Sampling combustible materials

Combustion or the presence of potentially combustible materials can have serious consequences and should always be fully investigated. Where combustibility is suspected bulk ground samples should be taken to depths as detailed in Section 16.1. Large-scale samples of up to several hundred kilograms may be required.

Laboratory tests for liability to combustion are noted in Section 17.5.5.

16.6 Sampling slags and other expansive fills

Reactions within the body of a fill containing slags, or colliery spoil or other pyritic materials can occasionally lead to damaging expansion beneath a building (see Section 11.4). Large samples, over the full depth and representing the fill mass, are needed.

Expansion phenomena are often accompanied by generation of sulphate/acid, with potential for aggressive attack on building materials. Analyses on samples taken under this heading should also test for this.

16.7 Sampling radioactivity

Sampling of radioactivity must only be carried out by experts or personnel trained under the supervision of the NRPB (see Section 23.3.3).

17. Detailed site appraisal

Section 6.1 outlines the general procedure for appraisal of the suitability of a site for a specified end use and means for obtaining suitability. Before continuing with the detailed practical application of appraisal procedure to specific dereliction hazards, it is necessary first to examine in greater depth two concepts introduced in Section 6:

* acceptance criteria   (Section 6.2)
* options for action   (Section 6.3)

17.1 Acceptance criteria

Current practice recognises four distinct types of acceptance criteria against which to judge the measured intensities of dereliction hazards:

Type
 A   Hazard unacceptable at any level
 B   Acceptance level defined by statutory authority
 C   Acceptance level defined by standard medical practice
 D   Acceptance level defined by standard technical practice.

Type A criteria: Hazards unacceptable at any level
Certain hazards, such as putrescent gas-producing material, spontaneously ignitable material, as well as substances such as asbestos and dioxin, etc., are considered unacceptable threats to specific targets under specific conditions at any measurable intensity. The decision here is simple: an unacceptable hazard must be remedied or the project abandoned.

Such a decision must, however, establish that the presence of the hazard constitutes a threat to the specific target. Thus asbestos exposed as dust in the atmosphere is unacceptably dangerous to site workers and others, whereas asbestos buried at sufficient depth may not constitute a hazard.

Type B criteria: Acceptance level defined by statutory authorities
For such matters as gas migration, leachate control and protection of water resources and the environment, statutory authority requirements may take precedence over technical considerations. Requirements may vary somewhat from one authority to another.

Type C criteria: Acceptance level defined by current medical practice
These are used for hazards affecting health. The list of acceptance levels is far from complete, and such data as exist do not always have clear cut-off values. In the UK the 'trigger values' (where available) proposed by ICRCL (1987) should be used to the exclusion of all other so-called standards. In making decisions the engineer will need advice from the chemist, microbiologist, medical authority or other specialists.

Type D criteria: Acceptance level defined by current technical practice
In some cases (e.g. sulphate attack on concrete) numerical criteria exist, derived from experimentally measurable effects of a given intensity of hazard on a specific target. In others engineering practice has established empirical relations, such as the tolerance of structures to settlement-induced distortion. However, in many cases (such as potential combustibility) precise numerical acceptance levels cannot yet be stated with certainty.

Limited range of available criteria
The dependable criteria so far available fall far short of a complete list of numerical values for all circumstances, and considerable judgement is needed in making decisions, especially regarding the environment. For the protection of groundwater resources and some other aspects of the environment, control is vested in the local authority, but no numerical criteria exist for the protection of flora and fauna. In the appraisal of health hazard, the ICRCL trigger concentrations have been established for only a very restricted range of substances.

17.2 Options and decision making

Section 6.3 listed the options available for the mitigation of incompatibility between discovered hazards and a proposed end use:

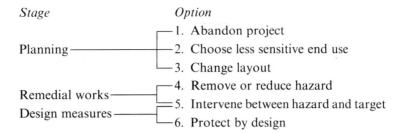

Stage	Option
Planning	1. Abandon project
	2. Choose less sensitive end use
	3. Change layout
Remedial works	4. Remove or reduce hazard
Design measures	5. Intervene between hazard and target
	6. Protect by design

Decision making exploits these options so as to arrive at the best development for a given site. If the planning options are unacceptable or insufficient by themselves, remedial works may be undertaken to increase the acceptability of the site, while design measures are intended to bridge any remaining incompatibility. Intervention between hazard and target (option 5) may have to be considered under two headings. Thus barriers against migrating contaminant may be regarded as remedial works but should also be taken into account in determining the suitability of design measures, whereas the provision of protective coatings against aggressive attack may be considered as a design measure.

17.2.1 Planning options

If the hazards are such as to render the proposed end use unrealisable under the existing site conditions, but the project is not to be abandoned, two planning options are open to the developer:

Option 2. Choose less sensitive end use
Option 3. Change layout.

Option 2. The engineer should keep the developer continually aware of the need for flexibility in planning. If, for example, the originally desired end use requires remedial measures beyond the developer's budget, the engineer should be prepared to recommend a change of end use in favour of one less sensitive to site conditions.

In inner-city areas with inherent high land values, however, that choice may no longer exist for two main reasons: first, the long time required to realise a development; and second, the urgent pressure on land in such areas. This means that definitive plans are often made before the true state of the land is accurately determined. Once the need for a given land use has been decided, there are limited sites suitable for its location. Given the high cost of land today, it is important to appreciate that a more expensive and rigorous remedial work may prove justified if it permits the realisation of a more sensitive end use and, thus, enhances the value of the site.

Option 3. The distribution and intensity of a hazard may be sufficiently variable, and the site of sufficient size, to permit adjustment in the layout of the proposed development. For example, the more seriously contaminated parts of a site may be allocated to uses more tolerant of the contaminant, such as roadways, vehicle parks, etc. The remaining areas can be used for constructing the planned buildings. Some precautions may be needed for protecting foundations, drains and water supplies, and to prevent lateral migration of contaminant. These measures will almost certainly be cheaper and less extensive than if the whole site were to be treated as seriously contaminated.

17.2.2 Remedial works and design measures

Remedial treatments for improving the suitability of a site involve the following options:

Option 4. Remove or reduce hazard
Option 5. Intervene between hazard and target.

These are dealt with in detail in Sections 18 and 19. The engineer should impress on the developer the need for adequate remedial works, and ensure that a realistic allowance is made in the budget. The need for adequate funds for such works is not always appreciated. Developments undertaken without properly planned remedial works have led to serious embarrassment on several occasions over the past decade or so. At Love Canal, USA, development over 1.4 ha of an unreclaimed dump of toxic waste drums resulted in a leakage of contaminants and damage estimated at \$760 million in 1978. At Lekkerkerk in Holland 270 houses, built in 1970 without prior remedial works on a contaminated site, were evacuated in 1978 with consequent damage estimated at 150 million guilders. In Britain, development at Thamesmead had to be suspended for considerable time upon the uncovering of previously undetected contamination, and at Willow Tree Lane, Hillingdon, a complete revision of plan was necessary after considerable sums had been spent on a development plan for 1600 houses, schools, etc., which proved unrealisable.

Design measures for foundations, services and structures (option 6) to satisfy the demands of an end use under the conditions remaining on a suitably treated site are dealt with in Section 20. Other things being equal, a combination of remedial works and design measures is usually less costly than a solution relying on defensive design alone.

17.3 Appraisal of fill

17.3.1 Factors affecting building development on fill

Building development on fill requires that:

- The fill be capable of sustaining the structural load without excessive movement, and
- The foundations be so matched to the fill and the overlying structure as to reduce to acceptable values all distortions transmitted into the structure as a result of differential movements within the fill.

The appraisal has to determine whether the fill/foundation complex will satisfy the requirements or, failing that, what would be the optimum combination of fill improvement and foundation design.

Other fill characteristics may give rise to additional problems: health hazards; combustion potential; aggressive attack and corrosion; expansion potential; gas emission and leachate production. The relative importance of the diverse hazards should be carefully weighed. For example, methane emission may prove a greater handicap than the presence of contaminants in the fill, and may even be a more intractable constraint than settlement.

This section deals exclusively with the consequences of the settlement potential of fills. The appraisal of the additional problems outlined above is covered separately in Sections 17.4 to 17.9. Charles (1984) has dealt comprehensively with the subject of fills. Reference may also be made to the BRE Digests 274 (1983), 275 (1983) and 276 (1983).

17.3.2 Appraisal procedure

Type D criteria. The suitability of a particular design of foundation and building depends on the tolerance of the superstructure to the distortions likely to result from expected differential settlements. Studies by several workers (analysed by Padfield and Sharrock, 1983) have established engineering criteria for acceptable distortions in framed and brittle structures. These criteria are the basis of the design principles adopted in Section 20.1.

Procedure. Appraisal is concerned with: the determination of the potential total settlement (and hence the differential settlement, Section 17.3.9); methods of combating the settlement; and the selection of a foundation system suitable for the desired building. Table 26 is a preliminary guide to foundation systems appropriate to various types of building on differing depths and conditions of fill.

Appraisal of fill thus proceeds in stages:

1. Determination of probable future settlement: for fill material this evaluation is critical and is dealt with in some detail in Sections 17.3.3–17.3.9.
2. Decisions on exercise of available options if computed settlement proves excessive (Section 17.3.10).
3. Decisions on exercise of countermeasures:
   —remedial treatment of fill to reduce settlements to acceptable dimensions (Section 18)
   —design measures for foundations and superstructure to combat distortions due to the residual potential for differential settlement in the treated fill (Section 20.1).

In practice, remedial treatments and design measures are studied together so as to determine the best combination of fill improvement and defensive design. For foundations bearing directly on shallow fills, remedial treatment may generally be cheaper than measures depending solely on sophisticated foundation design. For deeper fills the question is more open.

Classification: Fill treatment and residual settlement. Nearly all fills, which have not received adequate systematic compaction with placement in thin layers, will be found to require some form of remedial treatment before building can be contemplated. With some fills, in fact, no amount of treatment will render them suitable for building.

The principal aim of appraisal is to determine the likely behaviour of a fill and the advisability of remedial measures. To that end, a fill may be conveniently classified into one of three grades (after Charles and Burland, 1982), according to the feasibility, or otherwise, of remedial treatment:

Grade	Treatment
I	Unnecessary
II	Advisable
III	Of limited efficacy and possibly not economically feasible.

Most derelict fills today fall into Grades II or III.

This classification, in fact, is related to the overall settlement to be expected after the construction of the building (Section 17.3.8), and the main object of fill appraisal is the prediction of this settlement. This exercise demands sound judgement and a considerable experience of fills, as opposed to natural soils, on the part of the responsible geotechnical engineer.

Given the special problems of fill materials, their extreme heterogeneity, and the frequently large size of their discrete components, fills are rarely amenable to analysis from the results of small-scale laboratory tests. Today, however, several indirect approaches are possible (outlined in Sections 17.3.5 and 17.3.6) through the use of data derived from large-scale field trials by several workers over the past three decades.

Even with natural soils, settlement analysis may only be accurate to $\pm 50\%$, while fill is still less amenable to soil mechanics theory. In consequence, even an experienced geotechnical engineer can derive only an approximation to the overall movement. Nevertheless, experience shows that the results can be sufficiently close for classification purposes.

Building distortion tolerance. A building is at risk principally from differential movements, whereas overall uniform settlement does not normally cause distortion (Figure 27). Differential settlement may result from inherent non-homogeneity, variable depth, or other factors which render its accurate evaluation impossible. However, sufficient data exist to establish an empirical relation between differential and overall settlements (see Section 17.3.9). Acceptable degrees of building distortion are discussed in Section 20.1.

17.3.3 Factors affecting settlement

Important features of a fill site, which must be identified before an adequate appraisal can be made, include:

- Age of fill
- Depth of fill*
- Nature and type of fill*
- Particle size distribution*
- State of compaction*
- Extent of fill site
- Water table
- Nature of underlying formation.

These features affect the rate and extent of settlement, and thus have a determining influence on the classification of a fill, decisions on remedial measures, and the design of

Figure 27 *Garages distorted by settlement of fill (photograph by courtesy of GLC)*

foundations. Asterisked items are also important in the selection of parameters for the theoretical methods of Sections 17.3.5 and 17.3.6.

Age: Inert fills. For young inert fills (generally less than 5–10 years old), even when subject to building loads, self-weight creep remains the dominant component in the overall settlement potential, except for shallow fills less than 3–4 m deep. If the age of the fill can be determined, then the self-weight component can be assessed with reasonable confidence by the methods outlined in Section 17.3.5. Creep diminishes exponentially with time so that the major part is exhausted in the earliest years (often by more than half in the first year). The potential settlements likely to affect the structure are thus sensitive to age, with, typically, significant reduction in the potential between 2 years and 5 years.

Age: Biodegradable fills. Age is also significant in the appraisal of wastefills containing biodegradable materials. The gradual increase in the putrescible content of domestic wastes since the early 1940s (Figure 6) corresponds to increased gassing and combustibility hazards, and inferior density and engineering properties. Settlement of such fills results from a combination of normal creep and the effects of decay. Furthermore, uncontrolled recent wastefills with abundant putrescible content are scarcely amenable to remedial treatment by compaction techniques.

Table 16 Typical potential self-weight settlement of fill materials

Material	Potential self-weight settlement (as percentage of depth of fill)	Reference
Well-compacted, well-graded sand and gravel	0.5	Meyerhof (1951)
Well-compacted shale, chalk and rockfill	0.5	Tomlinson (1963)
Medium-compacted rockfill	1	Meyerhof (1951)
Well-compacted clay	0.5	Tomlinson (1963)
Lightly compacted clay and chalk	1.5	Meyerhof (1951)
Lightly compacted clay placed in deep layers	1 to 2	Tomlinson (1963)
Nominally compacted opencast backfill	1.2	Kilkenny (1968) Leigh and Rainbow (1979)
Uncompacted sand	3.5	Meyerhof (1951)
Poorly compacted chalk	1	Tomlinson (1963)
Uncompacted (pumped) clay	12	Meyerhof (1951)
Well-compacted mixed refuse	30	Meyerhof (1951)
Well-controlled domestic refuse placed in layers and well compacted	10	Tomlinson (1963)
Opencast backfill compacted by scrapers	0.6–0.8 (derived values)	Knipe (1979a)

Table 17 Compressibility of fills (after BRE, 1983a)

Fill type	Compressibility	Typical values of constrained compressibility modulus, D (MN/m^2)
Dense well-graded sand and gravel	Very low	40
Dense well-graded sandstone rockfill	Low	15
Loose well-graded sand and gravel	Medium	4
Old urban fill	Medium	4
Uncompacted stiff clay fill above water table	Medium	4
Loose well-graded sandstone rockfill	High	2
Old domestic refuse	High	1–2
Recent domestic refuse	Very high	<1

With older wastefills there is more probability that biodegradation will have been largely completed, and that the self-weight creep will have diminished to acceptable proportions. However, careful appraisal is necessary before any decision is made to build on wastefills less than about 30 years old.

Depth of fill. Fills may be approximately designated according to their depth, thus:

Shallow — less than 4 m deep
Medium— 4–15 m deep
Deep　 — more than 15 m deep.

Depth is important in that it controls the amount of overall self-weight movement to be expected at the surface of a given fill. However, for shallow and medium-depth fills, depth does not directly influence the technical feasibility of remedial treatment, since the classification of a fill for treatment depends on the settlement expressed as a percentage of the depth (but see Section 17.3.8 for reservations on classification applied to deep fills).

In younger fills depth is of serious importance in the design of foundations. This depends on the differential movements in the fill at foundation level, which are controlled by self-weight settlement.

Urban wastefills are usually shallow or of medium depth, although extensive in area. Opencast mining fills are practically always deep. Wastefills outside urban areas are generally in the medium to deep range. Generally, deep fills are more likely to be varied in composition, and to take longer to settle under self-weight and to cease gas generation.

Depth also affects the cost and choice of remedial measures and foundation design. In medium and deep fills the costs of all ground-improvement schemes will be closely related to the depth to which the fill is to be treated.

Nature and type of fill; particle size distribution. These give important qualitative information about a fill. Fills of natural soil should be classed as fine or coarse in accordance with BS5930, as this defines their behaviour. Coarse materials may be densified by compaction, in contrast to a consolidation process with fine materials. However, fills consisting of lumps of fine material in an unsaturated state behave somewhat as coarse materials.

Biodegradable materials, as well as some industrial wastes (such as coal residues), may be potentially combustible and thus liable to increased settlement. The quantity of biodegradable waste present also has a significant effect on settlement potential.

Degree of compaction. Good compaction can reduce the self-weight settlement potential by between 50% and 75%. Over the past 15 years or so a few opencast backfills have been placed in thin layers and rolled or otherwise laid by scraper to provide high-density, well-compacted material in the top 16 m. Such material may have adequately low settlement characteristics to permit building without remedial treatment.

In contrast, most other fills encountered on derelict sites will have been placed at low density, often by tipping with no compaction, or through water, or left to settle out from suspension in a lagoon. Such materials have poor engineering properties and high settlement potential.

Extent of fill site. Appraisal of differential settlement potential requires knowledge of the changing depths of the fill, together with the location and slopes of the fill site boundaries. On opencast backfills the position of the 'high wall' and intermediate steep changes of geometry are of particular importance. Even on sites where the general backfill mass may be of reasonably uniform density, the material round the boundaries may be distinctly inferior. This may result from the practice of tipping loose material down the sloping boundary faces, and the tendency of material to arch in contact with the slopes. In consequence, the boundary areas may exhibit large and unpredictable movements.

Decisions on remedial measures require knowledge of the volume of the fill, especially quantities to be removed.

Water table. Detailed knowledge of the present position and highest recorded previous levels of the water table, as well as forecasts, if any, of its potential permanent rise, are essential to the appraisal of the remaining potential collapse settlement of the fill.

17.3.4 Components of fill settlement

The vertical compression, to be considered in the classification of a fill for remedial treatment (Sections 17.3.2 and 17.3.8) and in the design of foundations (Section 17.3.10), is the total movement to be suffered by the building supported on the fill, comprising normally for inert, non-biodegradable fills:

- Residual self-weight compression (creep) of unloaded fill
- Primary compression of fill under structural load
- Additional creep due to structural loading
- Settlement of underlying formation.

If applicable, collapse settlement also has to be included.

The compression of the formation underlying the fill is amenable to conventional analysis and is not discussed further. However, the properties of the formation and its ability to carry the loads imposed by and through the fill must be demonstrated.

Prediction of the settlement potential of fill (by the methods set out in Sections 17.3.5 and 17.3.6) should be applied only to unsaturated fills devoid of biodegradable matter or, with caution, to biodegradable fills not less than 30 years old. Although data for biodegradable fills have been given in Tables 16 and 17 and Figures 7 and 8, the extreme heterogeneity of these fills of more recent origin dictates against settlement prediction, except for purely indicative purposes. Refuse fills have been known to take 30 years or more to stabilise. Monitoring may provide information on the creep potential of a given refuse fill, but can give no indication of future effects of biodegradation.

17.3.5 Prediction of self-weight compression

There are two possible approaches to the assessment of the residual self-weight creep of an unsaturated fill, differing in the source of data used:

(a) Extrapolation of monitored data obtained specifically for the given fill
    a.1. graphical
    a.2. analytical.
(b) Estimation from existing published data on similar type fills
    b.1 analytical
    b.2 graphical.

Both variants of method (a), as well as method (b.1) exploit the approximately linear relationship between settlement and logarithm of time elapsed since deposition of the fill. This is shown by the straight line on a semi-log plot (Figure 8). This relationship is also implicit in the exponential curves of Meyerhof (Figure 7) for the first ten years or so on a linear plot, which are the basis of method (b.2).

Mathematically, the vertical compression ($\Delta H/H$) of a thickness H of an unsaturated, inert fill between time t_1 and t_2 is given by:

$$\frac{\Delta H}{H} = \alpha[\log_{10}t_2 - \log_{10}t_1] \qquad (17.1)$$

where α is the compression occurring during one log cycle of time (e.g. between one year and ten years after placing of the fill).

Method (a) is the most reliable, but requires time for the monitoring operations. Method (b) relies on published data (α values, etc.) for other fills of similar type, and gives approximate answers without delay. The results are less dependable since the published data are rarely likely to apply exactly to a specific given fill. Preliminary estimates obtained by method (b) should be checked by monitoring. Observation shows that these methods usually give a slight overestimate after an extended period of years.

Method (a): extrapolation of monitored data. Given the approximately linear relationship referred to above, the monitoring of ground levels over an extended period (at least three, and preferably six, months (Figure 28)) may be extrapolated to predict the total compression of a free, unloaded, unsaturated fill settling under its own weight over an extended period. The extrapolation may be graphical (as Figure 28), or analytical by insertion in equation 17.1 of the appropriate α value derived from the monitoring results. The two procedures should give identical answers.

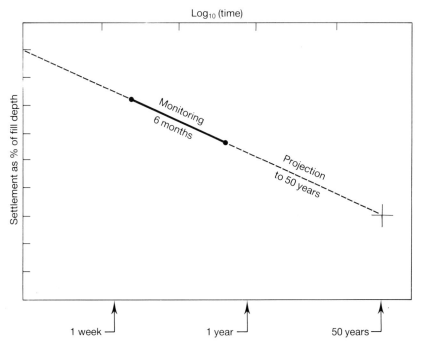

Figure 28 *Fifty-year self-weight creep settlement predicted from 6-month monitoring*

Method (b.1): analytical estimation from existing published data. Research at BRE and elsewhere has led to the publication of α values for certain types of fill (Table 18) for use in equation 17.1 (Charles, 1984). Given α and the age of the fill, the potential self-weight settlement (during the remaining life of a structure, for example) can be estimated.

Table 18 Creep settlement rate parameter, α (BRE, 1983a)

Fill type	Typical value of α (%)
Well-compacted sandstone rockfill	0.2
Uncompacted opencast mining backfill	0.5–1
Domestic refuse	2–10

So far, α values are available for only a restricted range of fill materials, and considerable judgement is required in selecting the value appropriate to a given site; the results should be regarded as indicative. The appropriate value of α may vary with the depth, nature and compaction of the fill, and will remain valid only so long a conditions within the fill (e.g. effective stress, moisture content, etc.) remain constant.

Method (b.2): graphical estimation from existing published data. For certain types of fill a rough approximation to the self-weight settlement can be derived from a knowledge of the behaviour of similar fills elsewhere. Field trials by various workers have established approximate data on the total self-weight settlement to be expected in various types and compactions of fill (see Table 16). On a linear plot, settlement follows approximately an exponential relationship to time, as established by Meyerhof (Figure 7). The type of fill and its age therefore determine the potential settlement remaining in a given fill. For suitable fills, the prediction of settlement by this method proceeds in steps:

- Determine fill type and corresponding total settlement (Table 16)
- Select curve most closely appropriate to type of fill (Figure 7). (It is found that the curves presented by Meyerhof for various types of fill coincide approximately when plotted to suitable scales)
- Assume this settlement practically complete in 10 years, not 5 as suggested by Meyerhof (see note). (This is simple to accommodate by altering the scale along the time axis)
- Apply the total expected settlement to the curve, so as to fix the settlement-axis scale
- Enter the curve at the point corresponding to the actual age of the fill, to obtain the value of the residual settlement.

Note: Recent field trials suggest that Meyerhof's original finding of virtual cessation of movement by the fifth year is somewhat optimistic, except for well-compacted fills, and that cessation by the tenth year is a more realistic assumption.

17.3.6 Prediction of compression under load

For fills devoid of biodegradable matter, and of sufficient age for the residual self-weight component of the settlement to be negligible, the settlement due to imposed loading can be assessed by conventional consolidation theory. For other types of unsaturated inert fill, two approaches are possible, as with self-weight compression:

(c) Extrapolation of monitored load test data obtained specifically for given fill
    c.1. graphical
    c.2. analytical.
(d) Analytical estimation from existing published data on similar type fills.

Monitoring (method c) is the more reliable, because it uses data obtained specifically for the given fill. However, it requires time and, for the assessment of deep fills, somewhat massive equipment (Figures 30 and 31).

In considerations of settlement under load, primary compression (ignored in self-weight settlement analysis) and residual creep are both important. Primary compression (Section 11.3.3) occurs rapidly upon application of the structural load. Provided the imposed loading does not bring the fill material to near bearing capacity failure, the primary compression can be predicted by the use of a constrained modulus D, such that:

$$D = \frac{\Delta\sigma_v}{\Delta\varepsilon_v} \tag{17.2}$$

where $\Delta\sigma_v$ is an increment of one-dimensional vertical stress due to the structural loading and $\Delta\varepsilon_v$ is that increment in vertical strain produced by $\Delta\sigma_v$.

Long-term creep is generally the more serious influence affecting the structure and is governed by the logarithmic relationship already established by equation 17.1.

Method (c): extrapolation of monitored load-test data. The primary compression under a building and the subsequent loaded creep can be extrapolated from an appropriate loading test. The extrapolation may be graphical (Figure 29), or analytical by insertion in equations 17.2 and 17.1 respectively of the values of D and α, derived from the test results.

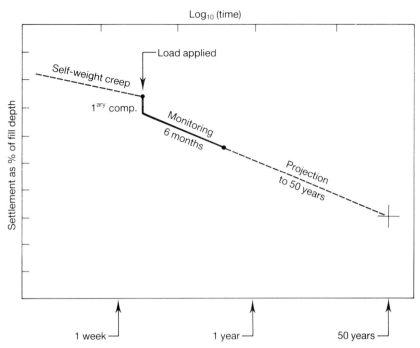

Figure 29 *Fifty-year building settlement predicted from 6-month load monitoring: shallow fill*

For the correct simulation of soil stressing the test load must be of dimensions and weight comparable to the definitive structural loading. Such a requirement can normally be satisfied without difficulty in practice for narrow strip footings which exert significant vertical stress down to about 1.5–2.5 m. For shallow fills, Charles and Driscoll (1981) have described a simple load test using a waste skip of bearing width and loading appropriate to a strip footing—such as might be suitable for light industrial buildings or for warehouses of large extent with independent floor slabs.

Figure 30 *Loading test on fill (photograph by courtesy of Cementation Piling and Foundations Ltd)*

Figure 31 *Loading test on a pad footing (photograph by courtesy of Cementation Piling and Foundations Ltd)*

For deep fills with massive raft foundations, a loading test becomes a more serious undertaking. However, for light development loads the settlements of younger deep fills are more influenced by the residual self-weight creep of the fill mass than by imposed loads. Extrapolated monitoring of self-weight ground surface movements may, therefore, be sufficient for the appraisal of a deep fill in cases of light development loading.

Method (d): analytical estimation from published data. Research at BRE and elsewhere has led to the publication of α and D values for certain types of fill for use in equations 17.1 and 17.2 respectively (Charles, 1984). Typical values of D are given in Table 17 for a range of fill types. They are applicable only to small increments in vertical stress (i.e. $\Delta\sigma_v \sim 100$ kN/m^2 imposed on an initial vertical stress of about 30 kN/m^2).

For long-term creep, Charles suggests that prediction can be made by using the appropriate α value, as in Section 17.3.5 (Table 18) with the proviso that day zero now corresponds to the application of the load. In the special case of suitably aged refuse fill (Section 17.3.4) the α value in Table 18 is not applicable. Instead Charles suggests a value of 1% as appropriate to the loaded state.

17.3.7 Collapse settlement

Likelihood of collapse. The possibility of future collapse upon inundation could be a deterrent to building construction which was to rely on the supporting capacity of the fill itself. Typically, collapse settlement may amount locally to 2–6% vertical compression, according to the type and state of the fill (i.e. sufficient to classify the fill as Grade III). Collapse is, however, not commonly a problem with properly compacted materials, particularly those of granular nature. The poorer the fill and its state of compaction, the greater the potential collapse movement to be expected under inundation. At Horsley opencast coal mining site Charles (1984) found pre-loading the backfill with a surcharge to be effective in reducing the subsequent collapse.

A rising water table can be the cause of collapse. This particularly affects opencast workings, pits, etc., which are usually dewatered during excavation and backfilling. Upon cessation of operations, the subsequent recovery of the water table should normally be complete within about a year. On larger sites, however, with deep pumping maintained locally, isolated areas may remain at risk for several years. Leigh and Rainbow (1979) and Smyth-Osbourne and Mizon (1984) have reported collapse damage due to a rising water table 24 years after the deposition of an opencast backfill (see also Sections 20.1.2 and 20.1.5).

Ideally, long-term monitoring of the groundwater regime is required before a reliable estimate of the likelihood of future collapse can be made. Failing that, it is at least essential to be sure that the water table is standing at the highest probable future level. Such assessment demands a detailed statistical study by qualified specialists of the local hydrology and hydrogeology.

Development in face of collapse risk. Decisions on whether to build on a fill which is still susceptible to collapse under unfavourable conditions must weigh the likelihood of future inundation against the urgency of the project and cost of precautionary measures. In some fills the local groundwater regime may be such that the fill may never be vulnerable to inundation from below.

Transfer of structural loading down to sound ground through piling is the most effective method of protecting a building from the effects of future collapse of the fill. The piles would have to be designed against the risk of overstressing from downdrag provoked by the collapsing fill. Additional measures to support ancillary structures and services entering the building would also be needed. Pre-loading with a surcharge, as above, may be an expedient, economic alternative.

The economic solution for a vulnerable fill is to delay construction until collapse can be verified as having taken place. If the groundwater regime reveals any risk of future rise in the groundwater above previous highest levels, the water table should be allowed to stabilise at its highest predictable level well in advance of any building work. The suitability of the fill can then be judged exclusively on the residual creep and the compression likely to be generated by the proposed building. These will probably be small in comparison with the potential collapse movement.

For opencast coal mining backfills, Leigh and Rainbow (1979) consider the first serious collapse settlement, consequent on the first major inundation of the fill, to be the most important. Provided other things were favourable, they would be prepared to build on opencast fill once the first major inundation had taken effect. Subsequent small fluctuations of the water table at a lower level are not considered to have serious consequences.

Even though the water table is predicted to remain at low level, the risk of collapse of the upper part of the fill through inundation from the surface should be considered. This is less predictable, although perhaps more controllable, than a rising water table. In fills with the upper layers already compacted, potential for collapse movements in these layers may have been reduced to acceptable dimensions. In untreated fills, densification, or excavation and proper compaction or replacement of the upper layers may be viable expedients. Care is needed, where stone columns are used for this purpose, to ensure that they do not form paths for water to penetrate to deeper, untreated levels. The alternative solutions intended to prevent surface ponding and infiltration through watershedding (convex surface slopes, surface sealing, ring drains, etc.) may not remain effective in perpetuity.

Charles (1984) has described a test to determine the susceptibility of a site to collapse settlement by flooding through surface trenches. However, it would be unwise to rely on single negative results from a trench water test because on other occasions the few recorded deliberate attempts to cause collapse (for ground-improvement purposes) by inundation from the surface have not been successful (Charles *et al.*, 1978).

17.3.8 Classification of fills for building

The need of remedial treatment, and the benefits obtainable, depend largely on the magnitude of the settlement likely to be suffered by the untreated fill under the superimposed load. The classification proposed in Section 17.3.2 can, therefore, now be defined in more detail (after Charles and Burland, 1982) as follows:

Grade I: Very small settlements—vertical compression nowhere more than 0.5%
Such fill forms good foundation material; treatment is normally unnecessary and there should be few problems (but see caveat on safety margins and foundation design on Grade I fills in Section 17.3.10).

This is typical of a granular fill placed under controlled conditions with adequate compaction. In practice, the only fills likely to be found approaching Grade I quality without further treatment are probably a few opencast mining backfills which have been earmarked in advance for building development. In some such cases compaction of the top 15 m or so may already have been carried out at extra cost. More rarely still, some fills of natural inert material may be found to have been deposited with stringent control to Grade I standard.

Grade II: Significant settlements—vertical compression between 0.5% and 2.0%
Ground treatment is desirable to convert the fill to Grade I material. Failing that, careful attention to the design of foundations is needed. This is typified by granular fills placed without compaction, with but little organic content and already in place for several years, and not liable to suffer collapse settlement. Opencast mining backfills usually fall into Grade II. Generally they are amenable to treatment to convert them into Grade I but, in practice, they are often left untreated and dedicated to agriculture.

Grade III: Very large settlements—vertical compression estimated to exceed 2.0%
Problems with settlement will be very severe. Ground-improvement techniques will be limited in what they can achieve, and the site may be prohibitively expensive to develop. Typical Grade III fills are recently placed domestic refuse with high organic content liable to decay and decomposition, or fine-grained materials transported in suspension and discharged into lagoons to form highly compressible cohesive fill, and fills liable to liquefaction. This grade will include some fills (opencast backfills, etc.) which are liable to, but have not yet undergone, major collapse settlement.

It is important to note that appraisal based on classification is not universally applicable; it should be used with caution in deep fills and fills of varying thickness (e.g. at quarry edges). However, in situations amenable to classification it permits appraisal through an assessment of potential settlement and, additionally, it offers the advantage of an approximate preliminary appraisal from readily observable fill characteristics.

17.3.9 Differential settlement

There are too many uncertainties for accurate prediction of differential settlement on fills. In this case, recourse should be made to the generally accepted rule in engineering practice that, in uniform ground, differential movement will not exceed 75% of the total overall settlements. Thus once the potential overall settlement has been estimated the likely order of differential movement can be assessed.

This is usually sufficient to cover likely local variations in a reasonably uniform fill. There are, however, particular situations where the risk of differential settlement is considerably higher. Such situations include:

• Abrupt changes of fill depth
• Steeply sloping surfaces of natural ground in contact with fill.

Thus the edges of old pits, quarries and mines should be avoided in the siting of buildings.

17.3.10 Options available

Defensive design for the protection of a building demands either the transfer of loading to sound ground through piling or the acceptance of some residual settlement, even after remedial treatment, with the load supported directly on the fill. In all cases the actions necessary will depend on the classification allocated to the fill.

Grade I fills. Given a sufficiently high level of control during densification works, proper compaction and diligence of supervision, a suitable fill uniformly densified to Grade I standard would form good foundation material, approaching greenfield quality. However, the achievement of overall uniformity to such a standard is difficult to demonstrate in the field. The uncertainty inherent in the nature of fill, no matter how well densified, calls for a greater margin of safety in design than is normal on a greenfield. Thus where direct bearing on fill is required, plain strip footings are usually not advisable on any type of fill. As a very arbitrary guide, on Grade I fills for small buildings for housing, the foundation type tends to depend on the fill depth, in general terms thus (see also Table 26):

Shallow	— reinforced strip footings
	— semi-rafts (as defined in Section 20.1.3)
Medium depth	— semi-rafts
	— rafts
Deep	— rafts.

Grade II fills. On these fills action will be needed to counter the settlement potential and decisions taken as to the best combination of available options:

Option 4. Removal of hazard (excavation and replacement with sound material) or
           Reduction of hazard (densification of fill to improve settlement characteristics)
Option 5. Intervention between hazard and target (piling)
Option 6. Design (special foundations, including rafts).

For light industry or housing, excavation and replacement is usually found to be attractive only for fills less than about 3 m in depth and above the water table.

For deeper fills the relative advantages of fill improvement compared to piling are somewhat site-specific. The mobilisation cost for vibro-compaction is often less than for conventional tripod-bored piles, and vibro-methods can be attractive for light industry or housing on depths to 6–8 m of fill.

For high-rise structures, large-span portals, cranes, heavy machines, silos and process plants on deep fill, the improvement potential of the fill itself is unlikely to be sufficient to permit direct founding on the fill, and piling is the logical solution. Piling is often a quicker operation than ground improvement, and may thus be attractive for low-rise structures on deep fill, even though it yields a better return on investment when used with high-rise or heavier structures.

Ground improvement, on the other hand, should be advantageous for low-rise buildings on rafts, lightly loaded framed structures, etc. on medium fills. On medium and deep fills treatment to partial depth may be attractive, but this requires full knowledge and understanding of the behaviour of the underlying, untreated fill material.

Of the ground-improvement methods available, dynamic compaction is appropriate when treating large areas, especially when the specific location of a building remains to be established. Vibro-methods may be used round the boundaries of such areas where the vibration and disturbance due to dynamic methods may be unacceptable. Foundations should not, however, span across fill areas subjected to differing types of remedial treatment (Section 18.5.1). On the other hand, for a small site or where the problem is to provide a foundation for a structure of specified location, vibro-methods will usually give a more satisfactory and economic solution. Vibro-methods and dynamic compaction are sometimes used in conjunction with each other to improve fill sites.

Grade III fills. On fills classified as Grade III, technical problems are likely to be severe and building development may be prohibitively expensive. Piling remains the only practicable option.

17.3.11 Landscaping

Given the long-term settlement potential of fill, restoration to the 'original contours' in flat areas may lead to waterlogging through failure of drainage systems. Good practice is to restore to a sufficient level above the surrounding land to provide surface drainage gradients away from the fill of not less than 3%. Such gradients should normally be acceptable for development involving roads.

17.4 Appraisal of health hazard

17.4.1 Concept of trigger concentrations

Warning on contaminant appraisal. Appraisal of the health hazard due to contamination is fraught with uncertainty, and demands considerable judgement and experience, and knowledge of ground chemistry. Current recommended good practice in the UK is set out in ICRCL Note 59/83 (2nd edition, 1987). The following notes are intended as a summary and guide to the ICRCL requirements, but no attempt should be made at appraisal until the ICRCL Note itself has been carefully studied and understood.

Type C criteria. Ideally, the appraisal of a potential health hazard, posed by a given contaminant to a particular target, should involve the comparison of measured levels of contamination against reference criteria established by current medical practice. Such an evaluation would require the prior specification of criteria having two bounding values:

- Lower (threshold) value
- Upper (ceiling or action) value.

Below the threshold, one could confidently ignore the presence of the contaminant, which would have insufficient power to harm the target. For concentrations above the ceiling, on the other hand, the target would be exposed to unacceptable risk, and remedial action would be mandatory before the project could continue.

The levels at which to set the bounding values for a given contamination would vary according to the target exposed to it. Thus zinc is known to be phytotoxic at concentrations well below the level at which it becomes dangerous to humans, whereas cadmium can accumulate in food crops to levels dangerous for human consumption, without apparent damage to the plant.

Trigger concentrations. In reality, statutory acceptance limits for contaminants on derelict land do not exist. Given the differing reactions of individual targets to specific stimuli, and the complexity of correlation between laboratory results and the interacting factors (ground chemistry/target sensitivity/target response), it is unlikely that universally acceptable values will be determinable experimentally.

Instead, ICRCL has put forward the concept of trigger concentrations, based on professional judgement of the existing information, and with account taken of the likely risks and costs of remedial works. At the time of writing, trigger values are available for only a few 'priority' chemicals, as given in Tables 19 and 20 (reproduced from ICRCL Note 59/83, 2nd edition, 1987). For the metals of Table 19, threshold values only are given; ceiling or action values for some metals are expected to be made available in a later edition of the ICRCL Note. Where both threshold and action values exist, the trigger concept divides the spectrum of contaminant concentration into three zones:

A   Concentration below threshold: site not hazardous, and no remedial action needed
B   Concentration between threshold and action level: site conditions progressively less then desirable. Careful appraisal necessary as to need for remedial measures
C   Concentration above action level: site conditions unacceptable and remedial measures essential.

In cases where only the threshold trigger value is available, corrective action may not become necessary immediately upon exceeding that value. Attention must, however, be increasingly paid to the potential hazard the more the trigger value is exceeded. Decisions in these cases as to the point at which to institute positive remedial action will require considerable expertise and the exercise of informed judgement on the relation of contamination to end use, rather than the application of rigid rules.

Table 19 Tentative 'trigger concentrations' for selected inorganic contaminants, from ICRCL Circular 59/83, 2nd edition, July 1987 (DoE, 1987)

Conditions
1. This table is invalid if reproduced without the conditions and footnotes.
2. All values are for concentrations determined on 'spot' samples based on an adequate site investigation carried out prior to development. They do not apply to analysis of averaged, bulked or composited samples, nor to sites which have already been developed. All proposed values are tentative.
3. The lower values in Group A are similar to the limits for metal content of sewage sludge applied to agricultural land. The values in Group B are those above which phytoxicity is possible.
4. If all sample values are below the threshold concentrations then the site may be regarded as uncontaminated as far as the hazards from these contaminants are concerned and development may proceed. Above these concentrations, remedial action may be needed or the form of development changed.

Contaminants	Planned uses	Trigger concentrations (mg/kg air-dried soil)	
		Threshold	Action
Group A: Contaminants which may pose hazards to health			
Arsenic	Domestic gardens, allotments	10	*
	Parks, playing fields, open space	40	*
Cadmium	Domestic gardens, allotments	3	*
	Parks, playing fields, open space	15	*
Chromium (hexavelant)[1]	Domestic gardens, allotments	25	*
	Parks, playing fields, open space		
Chromium (total)	Domestic gardens, allotments	600	*
	Parks, playing fields, open space	1000	*
Lead	Domestic gardens, allotments	500	*
	Parks, playing fields, open space	2000	*
Mercury	Domestic gardens, allotments	1	*
	Parks, playing fields, open space	20	*
Selenium	Domestic gardens, allotments	3	*
	Parks, playing fields, open space	6	*
Group B: Contaminants which are phytotoxic but not normally hazards to health			
Boron (water-soluble)[3]	Any uses where plants are to be grown[2,6]	3	*
Copper[4,5]	Any uses where plants are to be grown[2,6]	130	*
Nickel[4,5]	Any uses where plants are to be grown[2,6]	70	*
Zinc[4,5]	Any uses where plants are to be grown[2,6]	300	*

*Action concentrations will be specified in the next edition of ICRCL 59/83.
Notes: 1. Soluble hexavalent chromium extracted by 0.1 M HCl at 37°C; solution adjusted to pH 1.0 if alkaline substances present.
2. The soil pH value is assumed to be about 6.5 and should be maintained at this value. If the pH falls, the toxic effects and uptake of these elements will be increased.
3. Determined by standard ADAS method (soluble in hot water).
4. Total concentration (extractable by $HNO_3/HClO_4$).
5. The phytotoxic effects of copper, nickel and zinc may be additive. The trigger values given here are those applicable to the 'worst case': phytotoxic effects may occur at these concentrations in acid, sandy soils. In neutral or alkaline soils phytotoxic effects are unlikely at these concentrations.
6. Grass is more resistant to phytotoxic affects than are most other plants and its growth may not be adversely affected at these concentrations.

17.4.2 Trigger levels

In setting levels for trigger concentrations ICRCL has recognised three categories of contaminant/target (i.e. dose/effect) relationship:

- Contaminants hazardous at lowest measurable concentration (i.e. trigger value effectively zero). Examples are dioxin and airborne asbestos dust, but not buried asbestos
- Contaminants with definable dose/effect relationship (i.e. where a given dosage produces a determinable effect on target). Examples are phytotoxic metals (e.g. Zn, Cu, Ni as Table 19, Group B); phenol and organic compounds (contamination of water supplies, as Table 20); and cyanide (toxic through ingestion)
- Contaminants with no experimentally definable dose/effect relationship. Most contaminants affecting health fall within this category. There is at present insufficient evidence to specify precise trigger values for these contaminants, although Table 19 quotes threshold values for certain metals.

17.4.3 The use of trigger values in site appraisal

Trigger values have not yet been specified for many important contaminant/target relationships. The concept, however, offers a useful guide whose usefulness may be further

Table 20 Tentative 'trigger concentrations' for contaminants associated with former coal carbonisation sites, from ICRCL Circular 59/83, 2nd edition, July 1987 (DoE, 1987)

Conditions
1. This table is invalid if reproduced without the conditions and footnotes.
2. All values are for concentrations determined on 'spot' samples based on an adequate site investigation carried out prior to development. They do not apply to analysis of averaged, bulked or composited samples, nor to sites which have already been developed.
3. Many of these values are preliminary and will require regular undating. They should not be applied without reference to the current edition of the report *Problems Arising from the Development of Gas Works and Similar Sites* (Environmental Resources Ltd, 1987).
4. If all sample values are below the threshold concentrations then the site may be regarded as uncontaminated as far as the hazards from these contaminants are concerned and development may proceed. Above these concentrations, remedial action may be needed, especially if the contamination is still continuing. Above the action concentrations, remedial action will be required or the form of development changed.

Contaminants	Proposed uses	Trigger concentrations (mg/kg air-dried soil)	
		Threshold	**Action**
Polyaromatic hydrocarbons[1,2]	Domestic gardens, allotments, play areas	50	500
	Landscaped areas, buildings, hard cover	1000	10 000
Phenols	Domestic gardens, allotments	5	200
	Landscaped areas, buildings, hard cover	5	1000
Free cyanide	Domestic gardens, allotments, landscaped areas	25	500
	Buildings, hard cover	100	500
Complex cyanides	Domestic gardens, allotments	250	1000
	Landscaped areas	250	5000
	Buildings, hard cover	250	NL
Thiocyanate[2]	All proposed uses	50	NL
Sulphate	Domestic gardens, allotments, landscaped areas	2000	10 000
	Buildings[3]	2000[3]	50 000[3]
	Hard cover	2000	NL
Sulphide	All proposed uses	250	1000
Sulphur	All proposed uses	5000	20 000
Acidity (pH less than)	Domestic gardens, allotments, landscaped areas	pH5	pH3
	Buildings, hard cover	NL	NL

NL: No limit set as the contaminant does not pose a particular hazard for this use.
Notes: 1. Used here as a marker for coal tar, for analytical reasons. See *Problems Arising from the Redevelopment of Gasworks and Similar Sites.*
2. See *Problems Arising from the Redevelopment of Gasworks and Similar Sites* for details of analytical methods.
3. See also BRE Digest 250: Concrete in sulphate-bearing soils and groundwater (BRE, 1981).

significantly improved as a result of review, updating and re-issue by the responsible authorities. Provided they are regarded as guidelines and not as definitive standards, the ICRCL trigger values are the most realistic aid so far made available for contamination appraisal in the UK, and should be used to the exclusion of all other alternative classifications. There is serious risk of dangerous under-assessment, or costly over-assessment in the use of criteria other than ICRCL values for the particular conditions applying in the UK.

The trigger values to be used should always be those published in the most recent, updated issue of the ICRCL Note. The values are only applicable when used in accordance with the conditions specified in the ICRCL Note, and only after adequate site investigation. In using trigger values, the contamination of a site should always be judged on the maximum levels of concentration measured, and not on average values or results from analyses of mixed or composite samples.

Trigger concentrations are intended only for decisions on sites still awaiting development, they do not apply to sites already developed.

17.4.4 Appraisal expertise

The expertise necessary for decisions regarding health hazards will almost certainly be outside the ambit of the engineer and the geotechnical engineer. In this field the engineer will normally be obliged to rely on the advice of the chemist, who must have the necessary background experience to strike a balance between the twin evils of over-caution and over-optimism. The very high calibre required of the chemist, in terms of professional expertise and practical background of contaminated sites, has already been emphasised in Sections 14.1 and 15.7. It is in the realistic and objective appraisal of health hazards that those qualities will be most in demand.

Direct appraisal will be difficult for contaminants not yet included in the ICRCL guidelines. The chemist will have to use other sources of information such as foreign guidelines but bearing in mind that they may be based on data which may not be directly applicable to UK conditions (see Smith, in press).

On occasion, the appraisal of contamination may be relatively straightforward, for example where:

- No trigger values are exceeded
- Contamination levels are so obviously high as to remove any uncertainty about the need for remedial treatment.

However, even in cases such as these, the engineer would be ill-advised to act without recourse to the appropriate expertise.

17.5 Appraisal of combustibility

17.5.1 Appraisal procedure

Lack of dependable criteria. Combustibility appraisal is intended to determine whether material in or on a site will:

- Ignite, and
- Sustain combustion once ignited.

Appraisal depends on current professional practice, but reliable criteria have not yet been established. Combustibility cannot yet be confidently predicted by direct comparison of measurable properties (calorific value, etc.) against reliable established criteria. Appraisal requires the weighing of material properties against many other interacting factors, in which ambient conditions play a large part.

Intensive research is in hand worldwide but, at the time of writing, there are still no reliable numerical criteria which can be used in isolation. In consequence, combustibility appraisal requires experts with considerable experience in ground fire hazards and with a knowledge of laboratory testing of combustible materials.

Current practice. Despite the above handicaps, several expert practitioners have developed working methods in fire hazard appraisal, based on experience and the use of certain measurable properties as 'trigger' values. The following notes reflect current good practice, but they may become outdated once research provides more dependable criteria.

In the meantime, where a combustibility hazard is suspected, the advice and assistance of the Fire Research Station (FRS), Borehamwood, Herts (Tel: 081-953 6177) should be sought as early as possible at the planning stage.

In some reassurance for the current uncertainty, underground combustion is not a very common phenomenon on developable derelict sites.

17.5.2 Combustibles as contaminants

Underground combustion results from the oxidation of organic materials and carbonaceous minerals (oils and tars, solvents, coal residues, etc.) as well as non-carboniferous materials such as sulphur, zinc blende, iron pyrites and spent oxide in gas works residues. All of the foregoing, or their oxidation products, are contaminants or contain contaminants as troublesome impurities. Thus whenever combustible materials are found, one must be prepared to deal also with a contamination problem.

17.5.3 Definitions

A fire hazard is considered to exist whenever the site condition is such that smouldering can break out (and continue) in the ground mass, either spontaneously by self-ignition or triggered by external heating.

For this publication three categories of hazard are assumed:

A Spontaneously ignitable
B Combustible but not spontaneously ignitable
C Non-hazardous.

The present state of knowledge may not, however, be sufficient to determine in every case whether to allocate a combustible site to Category B rather than A. Environmental conditions may be just as important as calorific properties in determining the category of a site.

Category A: spontaneously ignitable or, more correctly, material in which the rate of self-heating (by slow oxidation, etc.) exceeds the rate of heat loss sufficiently to raise the temperature to a point at which self-sustaining smouldering becomes possible, given a supply of oxygen. This hazard must be treated before building can be contemplated; there is no alternative.

Category B: combustible but not spontaneously ignitable. Spontaneous ignition through self-heating is not possible, but the material may otherwise ignite through triggering by an external source of heat (surface fires, electric cables, etc.) or through migration of an underground fire from another area. The minimum treatment should prevent ignition temperatures from developing within the combustible.

Category C: non-hazardous because the combustible material is:

- Absent
- Too diluted to ignite
- Incapable of ignition (e.g. sufficiently compact to exclude air), or
- So protected (by cover layers, cut-off 'fire-fences', etc.) that its temperature can never rise to ignition point.

Opencast coal mining overburden by itself is considered to fall into Category C; any coal residues are very diluted, and no cases of fire hazard have been reported. However, underground coal spoils are notably richer in combustibles. Where these have been dumped alone or together with opencast overburden, combustion has sometimes occurred.

17.5.4 Assessment of hazard

With a fire hazard the issues are clear cut. The hazard must be eliminated or the project abandoned. If a fire hazard (Category A or B) cannot be disproved, treatment will be required. For Category B hazards design expedients may supplement the remedial treatment but are no substitute for it.

Uncertainty arises, however, from the difficulty in establishing whether a hazard does exist under the given condition of the materials in the ground. The parameters affecting the problem are numerous and varied, and a clear relationship often cannot be established between laboratory test results and potential *in-situ* ignitability/combustibility. In particular, there is no sure approach with current technology to derive reliable evidence as to whether a given material in the ground is capable of self-ignition (i.e. Category A).

In many cases the potential for combustion may be obvious from a history of fires on the site. In others the combustible content may be so low that no doubts arise. Between these two extremes of the combustibility spectrum a broad band of uncertainty exists. Research, currently in hand to define the factors affecting combustibility, may in future provide better criteria on which to predict the hazard (Beever, 1985).

17.5.5 Factors affecting appraisal

Important properties of combustibles, determinable from laboratory tests, are:

- Calorific value (C_v)
- Loss on ignition
- Flammable volatile content

but these have not proved infallible in predicting potential combustibility. Environmental factors also have a strong influence.

Calorific value. As a rough guide, ICRCL Note 61/84 (2nd edition, 1986) suggests a C_v of 10 MJ/kg (on air-dry sample, as BS 1016) as a possible fire hazard, while a C_v below 2 MJ/kg may not be dangerous. On the other hand, research at FRS shows that some materials of C_v well below 7 MJ/kg can propagate smouldering (ICRCL Note 61/84, 1st edition, 1984).

Until more dependable criteria are established, careful judgement is needed when using calorific values, which are not infallible indicators as to ignitability. Anthracite, for example, is a fuel with a high C_v which burns with little flame and high heat output, but requires raising to a high temperature under controlled conditions with high heat input before it can be made to ignite. In contrast, poor-grade coal wastes, with a C_v too low to be of commercial interest, may be more readily ignitable in air, and would thus pose a considerably greater combustibility hazard on a derelict site.

The most that one can say today is that samples with higher calorific values, lower ash and higher carbon/volatile contents may be more likely to be combustible than others.

Loss on ignition. This has been used as an indicator. Some practitioners quote a figure of 25% as a 'trigger' value, which has apparently been used with success in coal-mining areas. However, this empirical approach has evolved from long experience with specific coals, and its general application may be dangerous. If, however, the colliery of origin of the waste can be established, advice from a local mining engineer with experience of the particular coal may be invaluable.

Environmental conditions. Experience shows that spontaneously ignitable materials tend to ignite through self-heating sooner rather than later (i.e. in weeks rather than years). This implies that if a combustible, such as colliery spoil, has lain for an appreciable time after deposition without catching fire, it is unlikely to do so in the future, unless external conditions change, allowing drying out, for example (Beever, 1985).

Compaction also reduces the tendency to ignition through closure of air-access channels. In construction work it is thus considered generally safe practice to take loose colliery spoil (excluding washery waste) which has not caught fire in the past and to compact it for fill material. On the other hand, an underground fire, once started, tends to make its own air-access channels. It may therefore be useless to have compacted and uncompacted colliery wastes adjacent to each other in a fill if the loose material has any tendency to spontaneous ignition (Knipe, 1979b).

17.6 Appraisal of aggressive attack and corrosion hazard

17.6.1 Basic concepts

Conditions favouring potential attack on building materials, foundations and services may occur in industrially contaminated ground as well as in wastefill sites. Movement of corrosive chemicals in the ground may be slow and serious attack may not come to light for decades.

In general, the more soluble, reactive and/or mobile the chemical, the more rigorous must be the remedial treatment or design measures. Thus sulphates present as simple sodium or magnesium salts are more immediately dangerous than the low-solubility calcium sulphate (gypsum).

On the other hand, soluble salts may, under favourable conditions, be more rapidly leached away and become a less enduring threat. Thus, low-solubility calcium sulphate resulting from the oxidation of pyrites (Section 11.4.4) has been reported as remaining in the ground for up to fifteen years.

On greenfield sites, where the results from tests can be taken as representative of a stable ground chemistry, design is often the sole expedient adopted to safeguard a structural component against aggressive attack. The prime example is the specification of concrete in ground or near-neutral groundwaters containing sulphate.

On a derelict site, in contrast, the ground chemistry is often complex and dynamic. Reliance on design alone should, therefore, be used with caution, and only after the full potential for aggressive attack has been assessed.

17.6.2 Appraisal

Type D criteria. At present, the aggressivity to building materials of most substances commonly encountered on derelict sites is only known in qualitative terms. Typical qualitative data on aggressivity are given in CUR Report 31 (1965).

Appraisal must be on the basis of current professional practice but very few numerical data on acceptance criteria exist. In those rare cases where quantitative data are available, they usually refer to a restricted range of conditions. As an example, numerical values of acceptance levels relative to sulphate attack, quoted in BRE Digest 250 (1981), are only applicable to aqueous solutions at near neutrality, i.e. between pH 6 and pH 9.

Published dependable values for sulphate aggressivity in the usual acidic conditions of a contaminated site are not available at the time of writing (but see also Section 20.4 on recent new studies by Harrison, 1987).

Reference can usefully be made to CIRIA Report 98 (Barry, 1981) which deals with the subject of durability and protection of materials. However, the Foreword to that report states: 'There is insufficient knowledge at present on the levels of contamination which affect long-term durability and the time periods over which degradation mechanisms take place. For these reasons, the Report cannot be comprehensive or quantitative in its recommendations.'

Appraisal of the hazard presented by aggressive chemicals thus calls for considerable judgement by the engineer and practical experience on the part of the advisors. In all cases where potential aggressive attack is suspect, advice should be sought from corrosion experts or other appropriate specialists.

17.6.3 Decisions on remedial/design measures

Where investigation has revealed a potential aggressivity hazard, and where planning options 2 and 3 are either inappropriate or insufficient, possible further options comprise:

- *Remedial works*: ground treatment to reduce the hazard as much as reasonably achievable, by the removal, immobilisation or isolation (usually partial rather than complete) of the aggressive chemicals and by measures to control the groundwater regime (Section 19.7)
- *Design measures*: appropriate specification of materials, and detailing and construction for maximum practicable immunity to ground aggressivity, combined with protective measures applied to components in the ground (Section 20.4).

On derelict sites ground remedial works should always be carried out to the maximum extent practicable. Attempts to rely, instead, on purely defensive design measures only are usually more costly and less attractive, due to inherent uncertainties:

- Complex ground chemistry with aggressive concentrations beyond the resistance of normal materials
- Poor working conditions, not conducive to the good quality of construction necessary
- Difficulty of monitoring the continuing effectiveness of the protective system
- Unknown durability of protective materials, membranes, etc.
- Difficulty and cost of repair works.

In many cases, however, complete removal of a contaminant may be impracticable, and other of the foregoing remedial works must be carried out, combined with an appropriate degree of defensive design. By analogy, on a greenfield site it is not normally practicable to remove naturally occurring sulphate in the ground but, instead, to specify suitable concrete.

17.7 Appraisal of slags, mining spoils and building wastes used as fill

17.7.1 Recognition of problem

Type D criteria. Practically all steel slags as well as blast-furnace slags over 40 years old and sulphate-rich building wastes are 'undesirable' fill constituents under buildings (Section 15.10). Identification of these materials is fundamental in determining the suitability, or otherwise, of a fill.

For blast-furnace slag the acceptable sulphur levels are those in BS1047: 1983 originally intended for concrete aggregates and now widely applied for other uses. The standard excludes slags exceeding the specified levels of 0.7% for sulphate and 2.0% for total sulphur.

Where pyritic shales (especially those from formations listed by Taylor and Cripps (1984) Section 11.4) have been deposited on a future development site, it will be necessary to determine the likelihood of oxidation *in situ*, which may lead to damaging expansion and sulphate/acid formation. Nixon (1978) has given the following tentative criteria which may indicate a tendency to expansion:

- Appreciable percentage of pyrite in finely divided form
- Sufficient calcite to form gypsum. Nixon's studies suggest 0.5% CaO as a threshold value
- Total sulphur content (as SO_3) greater than acid-soluble sulphate content.

Sedimentary pyrite is said to be particularly prone to react expansively (Steward and Cripps, 1983). Where possible, examination of the parent formation *in situ* is advantageous, since the potential for damaging reactivity may not be evident unless pyrite is identified as the source material.

17.7.2. Decisions on protective measures

When an expansion hazard has been identified, the issues are clear cut. Building cannot proceed unless the hazard is treated.

All the materials noted as hazardous in Section 11.4 are undesirable for fill under buildings; the most certain protection would be to remove them completely. On the other hand, protective measures may occasionally be possible, as discussed later, thus:

- Defence against expansive hazards (Sections 18.6 and 20.6)
- Defence against sulphate/acid attack (Sections 19.7 and 20.4).

Where feasible, protective measures may be an acceptable alternative to the removal of these materials encountered *in situ*. The following notes should be taken into account in determining whether to apply protective measures.

Sulphate-bearing building rubble. Building wastes containing gypsum and other sulphates are undesirable fill in contact with concrete foundations or services, on account of potential attack. Analysed sulphate content is not usually a sufficient guide to the risk of attack, due to the great heterogeneity of building rubble. Isolated gypsum-rich masses or lumps may be encountered anywhere within the fill, whereas other sulphates may be disseminated throughout it. To avoid the risk of local concentrations of gypsum coming into contact with concrete, such building wastes should be avoided as fill. Alternatively, remedial treatment and design measures against sulphate/acid attack must be adequate to ensure no contact with sensitive building targets and no troublesome sulphate migration.

Metallurgical slags. Steel slags, as well as blast-furnace slags over 40 years old are undesirable fill materials due to their expansion potential. With old blast-furnace slags sulphate attack may also be a hazard. Where they are encountered in old slag banks, piling may be difficult to execute as an alternative to removal. Rafts on inert 'carpets' (Section

20.6) may be feasible, but design must take account of the potential total expansion. This will depend on:

- Depth of fill
- Percentage of expansive material ('free' lime, etc.)
- Percentage volume expansion potential (Section 11.4).

Pyritic materials. A few cases are known of damage to buildings by expansion of oxidising pyrite (Section 11.4.4), but there is scant documentation on protective measures. Prevention of oxidation would be an ideal solution, but almost impossible to maintain in perpetuity. One of the few reported proposals to that end (Grattan-Bellew and Eden, 1975) was to exclude air by maintaining a permanently high water table. In another case (Penne *et al.*, 1973) caustic potash solution was injected continuously into the pyritic ground, so as to create alkaline conditions unfavourable to pyrite-oxidising bacteria. Ground acidity was reduced from pH 3 to pH 7, at the expense of injecting 16 000 litres of water daily and the consumption of 12 tonnes of solid caustic potash over two years.

Such expedients are unlikely to be generally attractive, but it is worth noting that lowering the water table will increase the range over which chemical alteration is possible. In most cases it would be more practicable to accept the inevitability of oxidation, and either to exclude the pyritic materials or to take defensive measures. Such measures should primarily protect against sulphate attack (Table 21). For a few pyritic materials (e.g. those from

Table 21 Potential aggressivity from oxidation of pyrites

	Stage:	1 Fresh	2 Young	3 Acid	4 Oxidised	5 Old
Acidity	Actual	Low	Medium	High	Medium	Low
	Potential	High	High	Medium	Low	Low
Sulphate	Available	Low	Low	Medium	Medium	High
	Potential	High	High	High	High	High

formations listed by Taylor and Cripps, 1984) protection will also be needed against expansion hazard.

17.8 Appraisal of gas-emission hazard

This section is concerned with assessing the consequences of the methane content of landfill gas on a site and with decisions as to what measures may be necessary (or possible) to protect a desired end use.

The information given attempts to reflect recent practice and, to the extent possible, current opinions, but it may well become outdated by changing attitudes and increasing knowledge. An engineer concerned with building in the presence of gas should keep informed of eventual future developments and evolving recommendations and local authority requirements. Further useful information on gas-emission hazard can be obtained from Barry (1987), Crowhurst (1987), Pecksen (1985) and the Waste Management Paper series published by HMSO.

17.8.1 Type A Criteria

Landfill gas is principally a mixture of methane and carbon dioxide, derived from the anaerobic decomposition of organic materials. Serious sources of the gas may be: biodegradable domestic wastefill in old pits; sewage sludge deposited on land; river and dockland silt. Occasional sources of methane may also be: coal measures underlying a site, or having access through fissures; fractured gas mains; and, less troublesome, some natural peat deposits.

For a site to be intrinsically safe, methane concentrations must always be below an acceptable value. For example, BS6164 for safety in tunnelling requires explosion protection to equipment if methane concentrations cannot be kept consistently below 5% of the lower explosive limit (LEL) (i.e. 0.25% volume/volume in air). For buildings even lower values are often quoted (for example, 0.1% v/v in air—see Section 17.8.4). Such low values not only call into question the reliability of the available measuring methods but also are impossible to guarantee in perpetuity. Thus the 'trigger value' for methane is virtually zero.

The problem of appraisal is not, therefore, to determine whether an untreated site is safe *per se* but whether a cost-effective remedy can be found to render a desired end use safe.

17.8.2 Caution in building on gas-hazardous areas

Section 13.1.3 (*Inadvisable developments*) implies that, in very general terms, buildings should not be located on or adjacent to sites producing significant quantities of methane or carbon dioxide or containing significant quantities of putrescible material. At the present state of the art, that is the safest advice that can be given. There is, however, some divergence of opinion as to how to define 'significant', and how far and to what types of building this restriction should apply. Building development of gas-hazardous sites is a serious undertaking, requiring high technological inputs and long-term monitoring, and imposing legal responsibility and financial liability.

The fact remains that landfill gas can migrate long distances in permeable ground, and can be dangerous if it is able to accumulate in confined spaces at above the LEL for methane (i.e. above 5% v/v in air) or if carbon dioxide exceeds safe limits. Here, before progressing further, clarification demands an answer to a very searching question—can a dangerous concentration in fact build up in the confined space under consideration (Barry, 1987)? Landfill gas is a mixture of methane and carbon dioxide. Dilution often takes place during the migration of the gas, both underground and in the open. If dilution has progressed so much that the constituent gases are already below critical values before they enter a void, they cannot then accumulate to greater concentration, provided the atmosphere in the void is normal air. Conversely, the discovery of a significant concentration of a gas in a void space is evidence of yet higher concentrations in 'upstream' zones giving access to the void.

Technology is available today for the construction of safe buildings in gas-hazardous areas, but the funding required to finance the execution of properly engineered 'safe' designs may be so high that private houses are unlikely to provide an attractive return on investment. In contrast, warehouses, supermarkets and other large-volume enclosures of simple construction, which generate a continuous revenue, can be feasible propositions.

Such 'managed' enclosures can also more readily provide the necessary control of monitoring, and gas ventilation and dispersal, as well as checks on human activities likely to prejudice the integrity of gas defence measures. Where continuing safety depends on mechanical facilities (extractor fans, alarms, etc.) sufficient stand-by capacity must be available to allow for maintenance and breakdowns. This is more easily provided in large-volume enclosures than in private dwellings.

At the time of writing, attitudes appear to be hardening against any form of dwelling houses being built on gas-hazardous sites. Notwithstanding the available technology, houses with gardens should not be permitted because, with gardens, there is always the risk of inhabitants erecting sheds, greenhouses and other potential gas-traps.

17.8.3 Classification of sites

Degrees of risk. The site investigation for landfill gas should have recorded at the sampling points sufficient information for the production of maps (of the site and the relevant adjacent land) showing, month by month, for the monitored period (preferably at least six months):

- The extent of zones affected by methane
- Concentrations of methane, by volume in air
- Emission rates through the ground surface.

Reliable decisions cannot normally be made on anything less than this detailed information; isolated observations are certainly not enough.

As a rough guide, past experience permits the classification of the degree of risk in a methane-affected site, according to the percentage concentration of methane measured in boreholes (after Baker, 1987) as follows:

1. *Less than 0.1% methane.* Negligible risk to buildings, provided this concentration applies to the whole site; i.e. there must be no local areas of higher concentration at individual sampling points.
2. *Up to 10% methane.* Buildings will require remedial measures (typically, ventilated foundations).
3. *10–25% methane.* Risk substantial, and typically, buildings would require forced ventilation and alarm systems.
4. *Above 25% methane.* Building development not normally allowable. In case of over-riding practical or political necessity, consideration should be given only to institutional or industrial devlopments, with assured control of monitoring and dispersal and maintenance of systems.

For the higher concentrations, forced extraction of gas through pre-installed gas-gathering wells, piped gas-drains and pumping stations will be required before any form of building development can be contemplated.

To reiterate, however, gas concentrations on their own are not adequate for defining risk; gas volumes are also critical.

Caveat on low measured values. The above classification implies that, for sufficiently low measured values, the methane risk can be considered negligible. However, before consistently low values measured over an extended period are accepted as evidence of a safe site, certain questions should be asked (Barry, 1987):

1. Is low permeability in surface layers causing gas to migrate elsewhere?
2. Have probes been sunk to the base of the fill?
3. Have solid samples been analysed for degradable content?
4. Is there a low pH value or chemical 'poison' causing a suppression of methanogenic bacteria?
5. Is the fill deficient in moisture?
6. What CO_2 levels have been recorded?
7. Is the fill poorly insulated, causing aerobic conditions, and not permitting the ground temperature to rise?

The proportions of methane and carbon dioxide vary with the age of the fill, state of fermentation, and the presence of air, moisture, etc. Methane may be entirely absent in the early stages of biodegradation, or may be consumed in the case of underground burning, or by bacterial oxidation at low emission rates. The absence of methane from a borehole sample must not be taken to imply its absence in the future. The presence of carbon dioxide is usually a sign of underground biodegradation, which may start to emit methane at any time.

WMP 27 criteria. With the publication of Waste Management Paper 27 (HMIP, 1989) a different criterion of acceptability has been proposed. WMP 27 would require monitoring in gas-hazardous areas, unless gas production has fallen below the concentration level at which it constitutes a risk. This level is defined as less than 1% of flammable gas in the landfill gas and less than 0.5% by volume of carbon dioxide, measured in any monitoring point, taken on at least four separate occasions over a two-year period. WMP 27 would recommend agricultural after-use, until the wastes are so biologically stabilised that gas production no longer constitutes a risk, both on the site and for a 250-m wide band of adjacent land.

17.8.4 Generation and emission rates, and systems design

Gas generation. A surface emission of gas can be indicative of large volumes generated within the mass of a fill. *En route* to the surface from the seat of generation, methane tends to become attenuated, both by dilution by air and other gases in the ground and by secondary bio-processes (bacterial oxidation, etc.). Thus a measured concentration of 1000 ppm of methane at the surface may correspond to up to 70% at the source.

Gas-generation rate is a measure of the rate at which gas removed from the ground may be replenished. Gas may not always progress radially outwards equally in all directions. Ground geology, permeability, fissures, etc. may favour the transfer of greater volumes of gas in certain directions. Gas observations should, therefore, be sufficiently extensive to indicate the location of the seat(s) of generation. Together with gas-generation rates, this information is of importance in the sizing of active gas extraction and passive gas drainage systems (venting trenches, barriers, etc.).

Gas emission. In areas where measured gas concentrations are sufficiently serious to require protective measures, surface emission rates should be computed from the sampling point emission rates, measured as in Appendix 4.

The emission rate through the surface of the ground is defined as the volume flow per unit surface area in unit time (equivalent to m/s). It is an important factor in determining the potential rate of accumulation of gas in building foundations and other confined spaces above ground, and hence in the design of protective ventilation systems. It depends on the generation rate and the rate and direction of gas movement in the ground.

The emission from a given unit area of ground surface should theoretically be obtainable as the product of the measured borehole flow times the ratio of the given area (say, 1 m^2) to the cross-sectional area of the borehole. However, the measured emission at the sampling point is only an approximation to the emission rate under free-flowing conditions, and requires correcting assumptions and coefficients to suit a given case. On the basis of considerable practical work carried out by the Scientific Services Branch of GLC and Sir Frederick Snow and Partners, Pecksen (1985) has outlined working methods for deriving generation rates from first principles and accumulation rates in confined spaces from measured borehole emission rates.

In practical terms, methane emission studies at Surrey Docks resulted in the establishment of an emission rate of 0.3 l/min/m^2 as the level above which buildings would require designed anti-methane precautions (Thomson and Aldridge, 1983).

17.8.5 Dealing with gas

One essential for safe building development in a gas-hazardous area is to avoid accumulation of methane at critical concentrations in confined spaces. To that end, there are two separate problems to be addressed:

- *Primary protection*:   control of gas movement in the ground
- *Secondary protection*:   safe design of buildings, confined spaces, voids, manholes, services, etc.

Gas-movement control. This may be necessary for the protection of buildings (existing or proposed) on permeable lands adjacent to a gassing site as well as on the site itself. It may be the only remedy possible for existing buildings in cases where safe design modifications can no longer be incorporated. For projected new buildings it may usefully complement design measures to produce the preferred technical/economic solution.

Control involves various combinations of available technologies (see Section 19.5):

- *Active* pumped extraction of gas from the ground
- *Passive* interception of gas by deep, gravel-filled venting trenches which act as permeable walls, forming preferred paths for guiding migrating gas
- *Vertical gas barriers ('cut-offs')* for impeding the lateral mass transfer of gas
- *Horizontal barriers (covers)*: these can be effective in impeding mass surface emission into public open spaces, etc. (see also Section 19.1).

Pumped extraction is a serious and costly undertaking. For modern landfills (in consequence of the Control of Pollution Act 1974), it is usually an essential requirement of the licensing authority, and is the concern of waste management rather than a derelict site developer. For existing wastefills, on the other hand, Section 19.5 discusses the installation of pumped extraction systems in cases where reliance on passive venting trenches is not a sufficient safeguard. In such cases, installation and subsequent running (property management) costs may render building development an unattractive proposition.

Passive venting trenches can be effective in protecting developments on adjacent land. Adjacent land may be off-site or, in the case of sites with varying degrees of hazard intensity, areas away from the seat of gas generation. Venting trenches usually incorporate a low-permeability, vertical gas barrier on the side away from the source of gas to form an effective 'cut-off'.

Another use has been suggested for vertical gas barriers (Barry, 1987) i.e. around building developments on the less hazardous parts of a gassing landfill. In such cases the intensely gassing areas may be left for open space.

Horizontal barriers (low-permeability covers) are more suited to protecting open spaces rather than buildings, where they may be vulnerable to damage by settlement, pile penetration, etc. For buildings, in fact, there is often advantage in 'encouraging' gas to enter a properly designed underfloor void where it can be effectively dealt with. Horizontal barriers may cause problems by promoting lateral migration into other sensitive areas.

Safe design of buildings, etc. The basic objective of safe design should be to prevent dangerous concentrations of gas from reaching points within a building where it may come into contact with an ignition source. This involves two separate considerations:

1. Sealing all vulnerable entry points into the building proper
2. Channelling the gas into specially designed containment spaces, where it can be safely dealt with by evacuation, dispersal and dilution into the open air.

Practical realisation of these requirements is dealt with in more detail in Section 20.3. The 'safe' containment space is usually formed as an undercroft, beneath a building but separated from it by a gas-tight ground-floor slab. The floor of the undercroft is often formed of a permeable gravel layer to encourage gas migration into the void space. Evacuation requires ventilation, either natural or by an automatic forced system with detectors and alarms.

17.9 Appraisal of leachate

17.9.1 Statutory obligations

Leachate hazards. Infiltration of surface water or migrating groundwater into industrially contaminated sites, or fills containing putrescible domestic refuse or industrial wastes, can produce harmful leachate. A leachate-generating site constitutes hazards to several aspects of development, each subject to different appraisal criteria:

Target		Criteria type
Environment	B	Statutory constraint
Health	C	Medical practice
Buildings and services	D	Engineering practice

On a development site it is not normally feasible to remove the source of the hazard, due to the large volume of leachate-generating material involved. Instead, the preferred expedient generally is:

• Reduce the volume of leachate generated by controlling the quantity of water allowed to infiltrate, and
• Prevent entry of leachate into sensitive areas.

Environmental protection. Appraisal of hazards to health and aggressive attack on buildings/ services have already been dealt with elsewhere. Of equal importance, however, is the developer's obligation for the protection of the environment and third parties from the effects of migrating leachate. This gives rise to two fundamental problems:

1. Disposal of contaminated leachate; and
2. Control of its effects on ground and surface water regimes, within and beyond the confines of the site.

The statutory authorities accept that leachate, once formed, must be disposed of, but can impose restrictions on the volume and concentration of leachate discharged, whether by pumping into a foul sewer or surface waterway, transport for treatment or disposal, or dispersal by diffusion into adjacent geological formations. They may require to be satisfied as to the acceptance criteria to be adopted, the remedial works to be carried out and precautions to be exercised in perpetuity if necessary.

17.9.2 Appraisal procedure

Essential information. To fulfil the developer's obligation to the water authority an assessment is required of the leachate regime, covering plume shape and dimensions, flow rate and chemical composition, potential interaction with surface and groundwater resources including the main aquifer.

Such an assessment should be made by a qualified hydrogeologist with experience of contaminated ground and fill sites. Certain specialist consultants have developed mathematical modelling techniques, based on source/pathway/target analyses.

Mechanism of leachate dispersal. The route by which leachate leaves a site is strongly influenced by the hydrogeological setting. At disposal sites on geological formations of low permeability the rate of downward vertical migration may be less than that of leachate generation, and surface flows are to be expected. Conversely, at sites of high permeability the greater part of the leachate moves vertically beneath the site to recharge the local water table.

17.9.3 Decisions on remedial measures

Generally, leachate must be pumped, transported or allowed to discharge away from the seat of its formation. The engineer should ascertain the statutory authority's requirements on dilutions, discharge rates and treatment (if any), and attitudes to proposed remedial works. Consultation at an early stage may avoid lengthy delays in obtaining planning permission.

Where treatment is undertaken by the water company there will usually be a charge to the developer, who should also be advised as to continuing future obligations. Leachate generation is unlikely to stop with the development of the site and may continue for several years. Discharge into a sewer or watercourse will require continuing consent. The long-term open-ended financial liabilities involved should be carefully considered at the appraisal stage.

On permeable sites leachate, once formed, will tend to migrate through permeable strata. Remedial measures (Section 19.8) are concerned with (1) reducing the volume of leachate by impeding infiltration of water into leachate-generating matter; and (2) reducing contact between leachate and natural water resources.

This involves various combinations of covers and barrier walls, surface peripheral drains, surface contouring and other watershedding expedients, deep drains and control of the water table by pumping as necessary from wells, boreholes, etc. The water authority will require the residual leachate migration to be within acceptable limits of volume and concentration.

On sites of reduced permeability a large portion of the leachate may accumulate in surface hollows and must be discharged, pumped away or, in exceptional circumstances, removed by special transport. Discharge must be to water authority approval. Measures are also needed for reducing the quantity of leachate generated, as for a permeable site.

18. Remedial treatment of fill sites

This section deals with the conversion of Grade II fills (and, where appropriate, Grade III) to Grade I, as required for the safe realisation of buildings and services. (Section 17.3.8 gives definitions of fill grades.) This is achieved by improvement of the settlement characteristics, through the techniques outlined in Table 22. The specialist processes are well-proven engineering methods for the densification of loose and weak deposits, and are fully described in technical papers (for example: Greenwood and Kirsch, 1984; Greenwood and Thomson, 1984).

Table 22 Techniques for increasing the density of fill

Technique	Typical economic depth of treatment
Settlement under self-weight[1,2]	Full depth
Preloading	To about 15 m
Excavation and replacement or re-compaction[3]	Usually less than 3 m
Vibroflotation and construction of stone columns[2]	From 3 to about 6 m
Dynamic compaction[3]	To between 4 and 8 m

Notes: 1. Settlement under self-weight is a long-term process requiring acceptance of delay to the development programme.
2. The upper layers require further compaction.
3. Economic depth also limited by depth to groundwater.

Other hazards may also be present in fill and require treatment:

• Contamination
• Gas generation and emission
• Potential combustibility
• Aggressive attack on building materials
• Leachate production and migration
• Expansive reactions.

For convenience, these additional hazards are dealt with separately in Section 19. They should not be overlooked on fill sites. Gas emission, for example, may be an even more serious constraint than settlement.

The treated fill should be sufficiently improved that loaded areas settle uniformly without imparting significant tilt to superstructures. Between loaded areas, or between a loaded area and a service run, there may be appreciable differential settlement, which would have to be dealt with by careful design. It is also important to note that a structure on a treated fill may still be at risk from a poor compressible material underlying the fill.

18.1 Choice of technique

As shown in Table 22, fill can be improved either by free settlement under self-weight of the fill or, more rapidly, by the following techniques.

• Pre-loading
• Excavation and replacement
• Vibratory techniques, including stone columns
• Dynamic compaction.

The suitability of each technique is discussed under separate sections below.

Settlement under self-weight is essentially a long-term process. This option is feasible only so long as the necessary delay (until residual settlement potential has diminished to an acceptable value) can be accommodated in the development programme.

18.1.1 Relevant factors

The choice of technique depends on several factors:

- Type and nature of fill
- Depth of fill
- Water table
- Extent of site
- Economics
- Contamination (if present).

Possible interaction between contamination and treatment method may restrict the freedom of choice. For example, dynamic methods reduce pore volume, displacing gas and contaminated liquids. The need for continual removal of such liquids to avoid splashing, etc. may be an embarrassment to works organisations.

The simplest and most economic option is to treat the fill locally in the loaded areas under foundations, floors, services, roads, etc. This makes for quicker completion, less risk to site workers and third parties, and may be preferable where there is extensive contamination or gas-generation hazard. However, it does require close definition of development layouts at the design stage. This tends to limit the flexibility in future planning, whereas light industry (a preferred development on derelict land) may require up to 400% additional space available for future expansion: in other words, a high degree of planning flexibility. One alternative on larger sites accepts bulk excavation techniques on a large scale, which gives considerable flexibility in future planning.

The depth of treatment required depends on the type of structure to be supported; the depth affected by the treatment depends on the type of fill and method of densification. Dynamic methods compact to limited depth only; other methods can achieve deeper compaction, but economics usually dictate the depth compacted. The depths to which economic treatment can be effected by current techniques are indicated in Table 22.

Provided the fill contains no unsuitable materials, the density of shallow fills is usually improved by compacting it to the full depth, whereas in medium and deep fills it is not normally necessary to treat the full depth. An improved layer or 'crust' up to 6 m thick is usually sufficient to reduce the differential settlement to acceptable limits.

For small buildings on rafts on Grade II fills, densification of the upper 3–4 m may be sufficient. For larger units (or for small units on Grade III fills) the requisite depth may be considerably greater; densification may not even be feasible. In the latter case, it may prove economic to excavate and replace the top 5 m or so of the fill. If suitable, the excavated fill can be replaced in layers and compacted to the standard specified in Section 18.3. Unsuitable fill should be substituted with imported granular material. Rejected material may be taken to a dump off-site or reburied in a deep part of the site which is not to be developed for building. In either case the consent of the waste-disposal and other relevant authorities may be required.

18.2 Pre-loading

With this technique, the fill is compressed by a temporary loading designed to induce settlement prior to construction by producing stresses in the fill equal to, or preferably somewhat greater than, those to be expected from the future structural load. Costs are simple to estimate for this operation. Economy requires a locally available loading material, usually imported fill.

Pre-loading in strips, typically less than 10 m wide, causes immediate settlements, densifying the *in-situ* fill. The increase in density is greatest at shallow depths, with little change below 1.5 times the pre-load width.

Most fills are amenable to densification by pre-loading. However, the amount of settlement induced by the procedure depends on the pre-load, the nature of the fill, and the duration of loading.

With 'granular' fills settlement is fairly rapid, so that small areas can be treated at a time, and the load moved round the site. With 'cohesive' fills consolidation is slow but may be somewhat accelerated by vertical drains. Charles *et al.* (1978) found a 9-m surcharge superior to dynamic compaction (at comparable cost) in improving a 24-m deep clay fill.

In deep fills it is important to appreciate that self-weight settlement continues after removal of the pre-load. Pre-loading is considered by Knipe (1979b) to accelerate settlement. The amount of final settlement is unaffected, but acceptable values are reached sooner so that building construction can start earlier. Knipe found heave on removing a pre-load from a clay fill after more than one year (partly elastic rebound, but also a continuing swelling upon increase of moisture in the clay fill).

Pre-loading has been used on domestic wastefill at the Liverpool Garden Exhibition site where the areas designated for road construction were surcharged by 5 m of sand. The load was kept on each area for short periods and 300 mm of settlement was induced on a depth of 8 m. However, this example should not be taken to indicate that pre-loading can convert a domestic wastefill into a suitable foundation medium for buildings. The pre-loading has no effect on the rate or amount of settlement arising from biodegradation of the putrescible content of these fills.

18.3 Excavation and compaction or replacement

On large sites, where grading and landscaping may be required, the necessary bulk excavation can sometimes be economically and advantageously combined with ground-improvement work. The existing fill is excavated to a suitable depth and either re-compacted *in situ* or substituted by imported granular material placed in layers and adequately compacted. Placing and compaction should be in accordance with the recommendations of the Code of Practice for Earthworks (BS6031: 1981) and the Department of Transport's Specification for Highway Works (1986). The latter document and Section 19.9 give advice on the suitability of materials as imported fill. Geotextile sheets may be used to prevent the penetration of the fill into the underlying ground. Rejected fill material may be taken to a landfill dump or re-buried on a deep part of the site. In either case the consent of the waste-disposal and/or water authority may be required.

All surface water standing in ponds, hollows, etc. should be drained off and any soft surface material removed before filling starts. The gradients normally used for landscaping are not sufficient to cause stability problems but the increased surface run-off may cause troublesome erosion before the installation of the definitive drainage system. Suitable temporary collection systems should, therefore, be provided.

The potential disadvantages of the technique are:

- Excavation may disturb contamination
- Removal of fill from site puts extra traffic on local roads
- Fills of poor stability may require long batters to the sides of the excavation
- Where the original fill is to be re-used and compacted, considerable space will be needed for stockpiling. This expedient is thus less attractive on small sites.

Trench fill taken down to natural ground under foundations is a well-tried variant of this method in shallow fills. An economic example of this approach was a fill-replacement method reported by Wilde and Crook (1979) for houses and light industrial buildings on shallow fill (1–2 m). Foundation trenches were excavated under load-bearing walls down to sound ground and backfilled with hardcore (both imported and obtained from site-demolition works) as support to strip footings. The trenches were battered and wide enough for mechanical compaction plant. Strip footings were founded 700 mm below finished ground level and lightly reinforced wherever more than 600 mm of fill was replaced.

Weatherley (1979) reports a similar expedient, using pfa slurry in trench footings in combustible fill of colliery spoil. The pfa set hard and formed a continuous pier under the strip footings and incidentally, an effective fire fence.

18.4 Vibroflotation and vibrated stone columns

18.4.1 Available techniques

Nomenclature regarding vibratory techniques varies throughout the industry. The terminology adopted here is that proposed by the ICE (1987).

Vibroflotation/vibro-compaction by means of a vibratory poker was developed for the simple compaction of sand and granular fills having only minimal cohesive content. The technique has been extended to fills having sufficient cohesive content to inhibit compaction by vibration alone, by infill of the vibrated hole with small broken stone, compacted during the slow withdrawal of the poker.

Vibrated stone columns are constructed by vibro-replacement/vibro-displacement techniques for reinforcing weak soils and fills with properly compacted stone (typically, 20–75 mm size). Once the vibrator has penetrated to the required depth, a charge of stone is placed at the bottom of the whole. The vibrator is used to compact the charge of stone, and the surrounding ground if granular. By repetition, a dense stone column, tightly interlocked with the surrounding ground, is constructed to the surface (Figure 32). In practice, the great majority of vibratory treatments on derelict fills will involve vibrated stone columns.

Figure 32 *Stone columns in demolition debris fill (photograph by courtesy of Cementation Piling and Foundations Ltd)*

Under good conditions in granular materials a cylinder zone as much as 3 m in diameter can be compacted in a single penetration by plain vibroflotation, although a maximum of 2 m is more common. Generally, in cohesive fills the ground improvement is somewhat different, in that support for a substructure arises from the interaction of ground and vibrated stone column.

The vibrator generally weighs about 2 tonnes, plus the weight of extension tubes, and is typically suspended from a 20-tonne crane (see Figure 5). For small sites (less than 0.5 ha), vibratory techniques are less costly than dynamic compaction and can be used nearer to buildings and services. Relatively shallow treatment to 3–4 m depth at about 2 m centres has generally proved adequate for the majority of ground conditions encountered in the UK, although the equipment now available in this country is capable of treating most grounds down to about 10 m. In appropriate ground such treatment can be taken considerably deeper.

18.4.2 Methods

'Wet' and 'dry' processes have evolved, the choice depending on the nature of the fill:

Wet: unstable hole with high water table in granular fill, with soft clay content (possibly in layers). In this process high-pressure water jetting flushes out the soft clays, supports the hole, and assists the compaction of the stone.

Dry: stable hole with low water table in granular fill with no clay content. Compressed air jetting assists penetration and prevents collapse of the hole on withdrawal of the vibrator.

The wet process uses up to 25 m^3 of water per hour per rig, which in normal fills is discharged from the mouth of hole. Such a discharge may require special drainage arrangements and may not be acceptable to the drainage authority if contaminated by the contents of the fill. In consequence, in unstable fills there is today almost universal preference for a modified dry process with bottom feed. Stone infill is fed through the interior of the vibrator itself, which is slowly withdrawn in stages, maintaining the stability of the hole and compacting the stone as it is withdrawn. In unstable soils a recent trend is to form stone columns by means of a steel casing, thus ensuring a clean column without the use of a water flush.

The suitability or otherwise of various fill materials to treatment by vibratory techniques is indicated in Table 23.

18.4.3 Suitability of fills for vibratory treatment

Over half the annual expenditure on vibratory techniques is spent on the application of the methods to fill sites. They are appropriate to a wide range of fill materials, as indicated in Table 23.

Table 23 Suitability of various fills for vibratory treatment

Type of fill	Suitable	Remarks
Granular fill—sand, gravel, ash, slag, rubble, quarry waste	Yes	Subject to reservations in Section 17.7.2 on sulphate content
Cohesive fill—silty and clayey sands, silts, fine-grained fills	Yes	Undrained shear strength must be greater than 20 kN/m². Stone columns act as vertical drains
Mixed fill—mixtures of the above	Yes	
Unsaturated clay fill deposited for at least 8 years	Yes	Provided no danger of collapse on inundation
Fills with dispersed, small amounts of biodegradable material	Perhaps	Use conservative design approach. Biodegradable material should not have dimensions greater than a vibrated stone column diameter
Medium and deep fill	Perhaps	Partial depth columns do little to combat deep-seated self-weight movements
Soft clay fills	Perhaps	The process is generally inappropriate
Recently deposited fills (up to 10 years)	Unlikely	Self-weight movements are still significant
Fills liable to collapse upon inundation	No	
Biodegradable fill	No	Decay creates voids and loss of support for the stone columns. Columns provide leakage paths for landfill gas
Heavily obstructed fills	No	
Chemically unstable fills	No	
Potentially combustible fills	No	
Fills with soluble contaminants	No	Penetration may create danger to surface or ground-waters or create voids due to leaching
Heavily voided fills—with metal or plastic containers, concrete lintels, planks, doors, etc.	No	

Note: Limestone should not be used for stone columns where the pH of the ground is low.

Ideally, fills considered for improvement should contain no biodegradable refuse. Fills containing dispersed, small amounts of such refuse may occasionally be suitable for vibro-compaction. Such practice, however, involves some risk and in all such cases design should be conservative. Treatment, although accelerating self-weight settlement, has little effect on that component of such settlement which arises from biodegradation.

18.4.4 Specification

All densification works involving vibroflotation or vibrated stone columns should be in accordance with the Model Specification for ground treatment (ICE, 1987). One of the prime objectives is to achieve uniformity of the treated fills. To that end, good site investigation information is essential.

Good practice demands that the engineer be solely responsible for all pre-treatment site investigation, design of the foundation, supervision of the treatment, and specification of appropriate post-treatment testing procedures. In addition, spacings between holes should be clearly specified by the engineer. This calls for adequate knowledge on the part of the engineer regarding the characteristics of the machines to be used. Guidance Note NG 1.3.1. to the ICE Model Specification (ICE, 1987) also recommends pre-tender consultation between the engineer and reputable specialist contractors. The alternative, of specifying vibro-treatment solely by the performance of the finished works, is open to unequal tendering and may penalise the more experienced, conscientious contractor.

Under individual foundation bases a minimum of three stone columns should be adopted to ensure a stable foundation, less likely to be affected by differential settlement about any axis. Foundations other than bases are commonly reinforced strip footings or reinforced concrete rafts. In these cases lines of stone columns should be staggered.

In shallow fills, to about 4 m in depth, full-depth treatment is usual under dwelling-house foundations, and is also desirable for industrial buildings under both structure and floor. If economy dictates limited-depth treatment for industrial buildings, at least the stanchion bases should be founded on full-depth treated fill.

In deep fills (more than 15 m) the treated crust must have adequate stiffness (Section 18.1.1). Raft-type foundations should be adopted for dwellings whenever the upper layers only of a fill have been treated.

Testing of the degree of improvement is important, and zone tests or dummy footing tests should be included in the specification (Figure 33). The degree of improvement depends on the depth and type of fill and the energy imparted to it by the process. With present technology a granular Grade I fill may normally be expected to permit a bearing pressure in the range 100–250 kN/m^2 under properly designed foundations, with acceptable settlement. The corresponding bearing pressure on mixed cohesive/granular fills is in the range 100–150 kN/m^2.

18.4.5 Applications

In shallow fills with full-depth treatment to Grade I standard the treated ground will generally be found suitable for all applications within the bearing capacity ranges quoted above in Section 18.4.4. For deep fills with partial-depth treatment a somewhat more conservative approach is advisable. Loose, deep, natural sands have been successfully vibro-treated to permit the erection of high-rise housing and process plants, for example. Deep, treated fill materials, on the other hand, may possibly be found more suitable for low-rise development. Other applications in fill materials include foundations to storage tanks, and road and rail embankments.

Vibratory techniques do not compact the top 0.5 m or so of fill. This uncompacted material must either be removed or compacted by rolling, or the foundations must be taken through the uncompacted depth. It is also important to note that post-treatment density can be substantially reduced by drainage excavations, etc. The works should, therefore, be so planned as to locate post-treatment excavations at a safe distance from foundations and other critical areas.

Figure 33 *Dummy footing load test (photograph by courtesy of Cementation Piling and Foundations Ltd)*

In the development of inner-city areas, vibratory techniques have proved a successful expedient where old basements have been filled with demolition rubble. Such treatment has permitted the erection of new buildings on shallow foundations on these sites.

18.5 Dynamic compaction

18.5.1 Methods and effectiveness

In this process compaction is achieved by the repeated dropping of a heavy weight from a crane onto the fill surface (Figure 34). Typically, a weight of 15–20 tonnes is dropped from 15–20 m. The fill is treated on a regular grid of impact points, at a rate of about 5000 m²/week per rig in favourable conditions. Mobilisation costs are high so that the method is not competitive for sites where the area to be treated is, say, less than about 0.8 ha. Dynamic compaction can achieve bearing capacity and settlement performances similar to those for vibroflotation. At typical applied energies the process is not generally affected by obstructions, and tends to collapse voids and arching in the fill.

Before compaction is started, a free-draining, properly compacted, granular carpet (such as gravel or slag) is required over the whole site surface, especially on clay fills with standing water within the depth to be compacted. The carpet is generally about 0·5 m thick and may be one-fifth to one-third of the total cost of treatment.

The process should not be used within 30–40 m of existing buildings and services. The environmental disturbance is both physical and psychological, and can be severe. A possible solution is to use vibrated stone columns in the vicinity of buildings and services. However, it is not desirable to found a single monolithic structure on ground treated in part by dynamic compaction and in part by vibrated stone columns, on account of the likely differences in performance of the treated ground.

All densification works involving dynamic compaction should be in accordance with the ICE Model Specification for ground treatment (ICE, 1987).

Figure 34 *Dynamic compaction (photograph by courtesy of Keller Foundations Ltd)*

18.5.2 Depth of densification

In deep fills the depth to which the ground is affected corresponds roughly to

$$D = 0.5\sqrt{WH}$$

where D is densified depth (m)
       W is height of tamper (t)
       H is height of drop (m).

However, at typical applied energies with a 15-tonne tamper, the effective densification is found to be limited to the upper 6 m or so. For fills requiring densification down to only about 4 m a lightweight (8-tonne) tamper may be an attractive alternative, with height of drop and number of passes adjusted to suit the required energy input. Given suitable conditions, this may have great economic advantage, since the mobilisation costs for a lighter crane are halved compared with a 15-tonne tamper.

18.5.3 Suitability of fills for dynamic compaction

Dynamic compaction is most effective on coarse, free-draining granular material, and is especially suited to rockfill and gravel with boulders. Most unsaturated granular fills are improved by dynamic compaction, and some unsaturated fills of clay lumps respond fairly well to treatment. Some saturated granular fills may also be improved under careful control, but the effectiveness diminishes with decreasing permeability. Dynamic compaction may also reduce the residual creep of an unsaturated clay fill, whereas vibroflotation is virtually ineffective in this respect.

Dynamic compaction is a good method of compacting refuse and wastefills. Dynamic methods will not, however, eliminate biodegradation and, instead, may provoke or accelerate migration and/or emission of gas. If building is permitted on such a site, a possible compromise would be to pile the structures and to use dynamic compaction for ancillary areas of the site.

Dynamic compaction has little or no effect on saturated fine-grained cohesive soils, and this treatment is inappropriate for fills of such materials below the water table. Care is also needed in applying this technique to unsaturated clay fills, where heave may develop outside the crater. Tamping should be stopped and only resumed when the pore water pressure returns to normal. Deep drainage trenches filled with gravel may speed up the dissipation of pore pressure. All standing water on the surface and in craters should be pumped away, with care taken to record any discovered pollution.

Downie and Treharne (1979) describe the use of dynamic methods on 1.5 ha of a filled brickpit, containing waste (mainly industrial with organic material less than 10%) 10 m deep, for warehouse development. More recently, Thomson and Aldridge (1983) have described reclamation of the infilled Surrey Docks for housing development.

18.6 Remedial treatment of slags and other expansive materials in the ground

Expansion phenomena (Section 11.4) are often associated with sulphate and acid production. Protection against these hazards is considered in Sections 19.7 and 20·4.

Experience in the protection of buildings founded on pyritic and other potentially expansive materials is still somewhat limited. Section 17.7.2 outlines two ground treatments proposed in Canada, with the aim of creating conditions in which the damaging reactions cannot occur in the ground. These procedures would be very difficult to realise in perpetuity and, at present in the UK, are mainly of academic interest.

If these materials are *in situ* the ideal safeguard would be the complete removal of the hazardous material. On the other hand, provided adequate protection can be ensured by other means, it may occasionally be feasible to accept leaving the material in place. Defence of a building then relies mainly on design measures (Section 20.6), recognising that expansion is inevitable, sooner or later.

It is highly undesirable to import expansive material as fill under proposed new buildings. If no choice is possible, then adequate protection must be provided (Section 20.6). On the other hand, expansive materials may be acceptable as bulk infill, provided sulphate/acid migration will not be a problem, and, further, that future unforeseeable building developments can be adequately protected.

19. Remedial treatment of contaminated sites

This section outlines the two current treatments in the UK for protecting targets from contaminated ground:

1. Cover over contamination
2. Removal of contaminated material.

Brief mention is made of possible future developments in the destruction or neutralisation of contaminants, either *in situ* or on- or off-site Advice is also given on sources of clean replacement fill material and topsoil (Table 24).

Control of special hazards concerns, in addition:

* Gas emission
* Potential combustion
* Potential for aggressive attack
* Leachate generation.

Although dealt with in this section, these latter hazards are also important for fill sites.

As well as remedial treatments, the essential works listed in Section 7.4 will be required, either as routine preliminaries or obligatory for third-party safety.

19.1 Use of covers

19.1.1 Development of cover concept

Covering over contaminants left *in situ* is the first of the two reclamation techniques currently used in the UK. Covers were originally planned to keep surface and sub-surface targets from direct contact with contaminants, with scant regard for the need to protect groundwater or the main aquifer, and with little understanding of contaminant migration. Covers to fulfil such a limited specification could be of simple design, of clay or other fine-grained material, and fairly cheap compared to the alternative of removal or destruction of the contaminant.

Fine-grained covers of low permeability suffer, however, from upward migration of contaminant through soil suction in time of drought (Section 12.14) or through penetration by rodents, earthworms or roots. The rational solution was the introduction of a designed 'break-layer' of coarse granular material of low capillarity against soil suction (Roberts and Gemmell, 1979; Cairney, 1984, 1985, 1987), or a 'bio-barrier' of somewhat coarser material against penetration by earthworms, etc. (Cline *et al.* 1980). By themselves these granular-layers do little to protect the aquifer from downward-percolating contaminated groundwater. Break layers, however, have been used with success on old industrial sites in river valleys, where there is, fortunately, often sufficient thickness of saturated alluvial clay below the contaminant to safeguard the underlying aquifer. Figure 35 illustrates this particular line of evolution in cover concept over the past decade or so.

An aquifer remains vulnerable, however, wherever it is in contact with permeable strata accessible to contaminated groundwater. It is probable that the importance of the local aquifer has been insufficiently appreciated in industrial areas in the UK where much of the water supply arrives by aqueduct from distant mountains. There is now, however, growing pressure for the effective safeguarding of local ground and surface water resources, as well as the prospect of the introduction of more stringent EC regulations. For these conditions, low-permeability 'infiltration-barriers' combined with granular 'water-drainage' layers become important (see also Section 19.8).

Cover layers may give no protection where impregnations of tar (or other 'thick' liquids which do not disperse in the groundwater) remain sufficiently soft to be squeezed to the

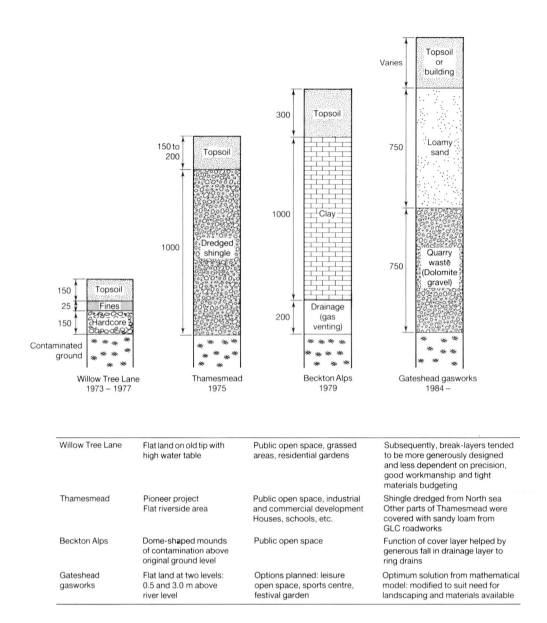

Willow Tree Lane	Flat land on old tip with high water table	Public open space, grassed areas, residential gardens	Subsequently, break-layers tended to be more generously designed and less dependent on precision, good workmanship and tight materials budgeting
Thamesmead	Pioneer project Flat riverside area	Public open space, industrial and commercial development Houses, schools, etc.	Shingle dredged from North sea Other parts of Thamesmead were covered with sandy loam from GLC roadworks
Beckton Alps	Dome-shaped mounds of contamination above original ground level	Public open space	Function of cover layer helped by generous fall in drainage layer to ring drains
Gateshead gasworks	Flat land at two levels: 0.5 and 3.0 m above river level	Options planned: leisure open space, sports centre, festival garden	Optimum solution from mathematical model: modified to suit need for landscaping and materials available

Figure 35 *Evolution of cover layers in UK practice*

surface. It is thus of paramount importance not to create new spillages by careless site working (compare Figure 18, page 68 and Figure 36). Tar spillages tend to lose their volatiles slowly and become more viscous with age, so that the ground may eventually become amenable again to reclamation by covering.

It is important to appreciate that a cover should be executed above the water table. Once the water table rises into the cover, component layers below the saturation level cease to be effective for their designed purpose. It is also perhaps important to note that covers are less readily accepted in the USA and on the continent of Europe, where they tend to be regarded as temporary measures.

On specific sites other problems may arise, such as the need to control gas migration and emission; exclude air and heat from combustible materials; prevent penetration by rodents, earthworms, plant roots, etc. Where covers are laid without removal of an equivalent thickness of existing material, the consequent raising of the surface level may be advantageous for watershedding. The additional loading may, however, provoke increased settlement.

Figure 36 *Pumping water and tar from an underground gas holder (photograph by courtesy of Environmental Safety Centre, Harwell Laboratory)*

19.1.2 Function of a cover

The logical result of the foregoing is the proper design of a cover to fulfil several, often conflicting, functions. It is unlikely that a single layer will satisfy all the requirements of a given site. This leads to the concept of a layered sandwich cover, discussed in detail by Parry and Bell (1985) and Cairney (1987). The design of a cover is a site-specific operation, requiring a full site investigation and a proper appreciation of the functions of each component layer.

Depending on the given site and type of target at risk, a cover has some or all of the following functions:

- Providing a non-flooding, non-eroding and non-toxic surface
- Supporting vegetation
- Supporting structure/traffic loads without unacceptable differential settlement
- Resisting penetration by rodents, earthworms or roots below selected depth
- Physically separating contaminant from direct contact with surface and sub-surface targets
- Protecting ground and surface water resources by limiting downard infiltration of surface water and rainfall into contaminant
- Controlling emission and migration of landfill gas
- Preventing upward migration of contaminant, through soil suction or upward excursion of groundwater
- Protecting underground combustibles from air access and heating by surface and sub-surface sources
- Neutralising contaminant.

In addition, a cover should not require precise workmanship for its realisation, should retain its integrity for the whole life of the project (say, 100 years minimum), and should permit commonplace excavations (e.g. for maintenance, horticulture, etc.) in the future without prejudice to its basic functions. Consequently, a cover should be generously designed, with no individual layer less than 100–150 mm thick (300 mm in the case of topsoil). There have been instances of over-precise design where special layers of only 25 mm were specified. The impracticability of executing such thin layers has been to the detriment of the future integrity of the whole cover.

19.1.3 Materials for cover layers

The various functions required of a cover lead to the concept of layers of different materials designed for specific purposes. In practice, several functions may be performed by one of the component layers, acting as a 'multi-purpose layer'. Some suitable materials for individual functions are listed below. For fuller details see Parry and Bell (1985) and Cairney (1987).

Function	Suitable material
Vegetation support	Uncompacted topsoil (to BS3882: 1965)
Load support	Properly compacted granular or other material of adequate shear resistance, as separately provided to suit other requirements
Anti-rodent (etc.) action	*Bio-barrier*: loose rock layer below topsoil with filter incorporated
Limit water infiltration	*Infiltration-barrier*: compacted clay or other fine-grained low-permeability material
Gas control	*Gas-barrier*: compacted damp clay or other low-permeability material, or membrane; with gas vents
	Gas-drainage: coarse granular material below gas barrier (see also Section 19.5.2)
Prevent soil suction	*Break-layer*: coarse granular material
Limit water table rise	*Water-drainage*: coarse material with field drains
Protect combustibles	*Insulation layer*: material already provided for other purposes (1 m min. thickness)
	Air-barrier: compacted damp clay (see also Section 19.6.5)
Neutralise contaminant	*Chemical barrier* (see Section 19.1.4)
Prevent loss of fines	*Filter layers* designed as required at boundaries of granular materials

Before definitive design can start, the materials proposed for a cover should be fully laboratory tested. Apart from load-bearing capacity and other soil mechanics aspects, important properties are permeability, capillarity, expansivity potential, chemical composition and reactivity with contaminant.

Granular materials should be inert, locally obtained if possible, not liable to crush or crumble, with low silt/clay fractions and, except for chemical barriers, not likely to react with contaminants.

Clay materials, when placed near the surface, are poor supporters of construction traffic in wet weather. They are liable to crack and lose their low permeability in time of drought unless placed sufficiently deep to remain always damp. Near an exposed surface they should be covered with topsoil. Clay is also attacked by some organic solvents, to the detriment of its low-permeability properties.

19.1.4 Factors affecting design of covers

The design of break-layers involves soil science principles somewhat outside normal soil mechanics practice. The special considerations necessary are touched on very briefly in Section 19.1.5 and are dealt with in detail by Cairney (1984, 1987). With that exception, the design of cover layers follows basic engineering principles. For each case the optimum remedy depends on climatic conditions, site hydrogeology, hazards present, the target at risk and the materials available.

The application of particular materials to fulfil the specific functions listed in Section 19.1.3 requires an appreciation of the factors discussed below.

Watershedding. Surface flooding should be avoided to limit infiltration. Expedients include peripheral interceptor drainage, surface compaction and landscaping to adequate convex slopes, internal drainage, topsoil and grass cover (see also Section 19.8.1).

Vegetation. Land not covered by buildings or other hard surfaces should be vegetated (shallow-rooted grass species, etc.) to protect from erosion, to improve soil stability and aesthetics, and remove surface water through evapotranspiration.

Load support. Where loads are to be supported, all materials selected for other purposes (except topsoil) should be suitable for proper compaction.

Bio-barrier. Earthworm or rodent boring or root penetration can cause upward transport of contaminant into sensitive topsoil, and can lead to breakdown of sealing layers. A layer of loose, poorly graded stone (40–80 mm size) with air voids, at required depth below topsoil, has successfully prevented such activity (Cline *et al.*, 1980).

Infiltration-barrier. This limits downward percolation into a contaminant. Properly compacted fine-grained soils, clay, etc., about 0.3–0.5 m thick, are suitable materials, provided clay is not allowed to dry out and crack.

Gas control layer. Horizontal layers may be buried across the site to control the mass upward movement of gas over a large area. They consist of a permeable 'gas-drainage' layer of poorly graded large broken stone, capped with a low-permeability 'gas-barrier' layer, vented at designed intervals. The gas-barrier layer may be of damp clay, provided the layer is deep enough in the ground not to dry out and crack. Alternatively, the barrier may be of synthetic membrane (e.g. 1000-gauge polythylene), protected above and below by buffer layers of 100–200 mm of sand or pfa. The gas-barrier has to be accurately sealed into gravel-filled peripheral cut-off trenches; proper venting is important. Careful detailing is needed where services penetrate the membrane.

Break-layer. This guards against soil-suction-induced contaminant migration during drought periods. It requires careful design, but about 0.5–1 m thickness of coarse granular material (quarry waste, etc., 25–50 mm size) below a sensitive target is usually adequate. A break-layer specifically designed against soil suction is required only for water-soluble contaminants, and only in cases where sensitive targets are less than 4 m above the highest possible contaminated water table in fine-grained material. For other situations a simpler capping to suit other requirements is adequate. Break-layers are not usually needed under buildings where the upper soil layers are not likely to dry out in drought.

Water-drainage layer. This can have two useful functions: (1) to limit the rise of the water table, and (2) where levels permit, to drain off surplus downward-percolating water which might otherwise infiltrate into the contaminant. It is of poorly graded stone with field drains along the lower face. The field drains should be sized to ensure that the water table will never rise above them. Water-drainage layers may usually be combined with break-layers.

Insulation-layer. Combustible material underground must be prevented from heating by an underground source (Section 12.9). One metre thickness of suitable material is usually a sufficient insulation. There is normally no need to provide a special layer, because the material provided for other purposes will usually be adequate.

Air barrier. Air must be excluded from combustible material. 150 mm of compacted clay may be adequate provided it stays sufficiently damp not to crack.

Chemical barriers. In scrapyards, etc. a limestone layer may be useful to correct acidity and to 'fix' metallic anions (chromates, etc.) and to form insoluble calcium salts. Clay also has the useful property of fixing heavy metal cations by ion exchange.
Note: chemical barriers should not be combined with break-layers; the latter should not react with contaminants.

Filter layers. These are required to prevent loss of fines at the boundaries of coarse granular layers. They may be of geotextile or designed graded material.

19.1.5 Design of multi-purpose layers

In order to show the specific remedy to combat each hazard the various functions of a cover have been considered separately in the previous section. In practice, it is rarely necessary to construct a sandwich of great complexity, because many diverse functions may be performed by a single multi-purpose layer. Judicious design will usually permit a granular layer to function both as a break-layer and as a water-drain, provided it is designed for the severest demands of both functions. Similarly, a single clay layer may function as an infiltration-barrier and as a gas-barrier.

As an example, a granular layer may be required to:

(a) Prevent rise of contaminated water table in winter (water drainage)
(b) Protect the aquifer by draining off surplus, downward-percolating, water before it reaches the contaminant (water drainage)
(c) Prevent soil suction in drought (break-layer).

In the UK climate, functions (a) and (b) usually predominate. Drought is a rarer occurrence but, unless it is anticipated, soil suction can, during the life of a project, subject a target to dangerous and increasing accumulations of migrated contaminant.

A corrective granular layer should, therefore, be designed to meet the objectives of (a) and (b) but with modification as necessary to provide a break-layer to satisfy function (c). Cairney (1984, 1987) has dealt in detail with the design of break-layers. The following notes are based on Cairney, but his publications should be studied whenever break-layer design is under serious consideration (Figure 37). Design proceeds in stages thus:

1. *Identification of most sensitive target*. For upward migration this may be a surface target, buried service, or deepest foundation or plant root, etc. For downward percolation it may be the local perched water table or the main aquifer
2. *Delimitation of groundwater excursions*. Herringbone field drains (laid at the bottom of the granular layer) should be sized for design rainstorm so that the water table never rises above design level. The drains also contribute largely to reducing percolation down into contaminant. Effluent is taken from above the contaminant level and is normally clean enough to be discharged into surface drains
3. *Determination of severest probable drought condition* and corresponding capillary rise above water table in the granular material. This is exclusively a soil-suction requirement (i.e. as function (c) above). Application of Darcy's law to suction head gives depth of granular layer to ensure that no upward capillary movement (i.e. contaminant) reaches the target. Mathematical models have been developed for the design of break-layers. One of the most used is by Bloemen (Cairney, 1985).

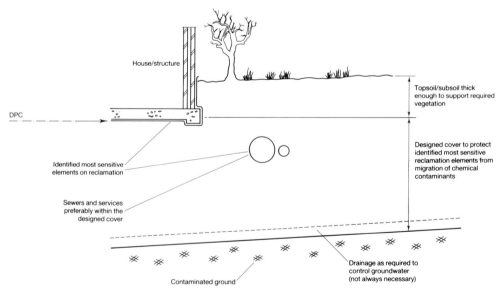

Figure 37 *Basics of cover layer reclamation (after Cairney, 1985)*

The granular layer may be supplemented by a low-permeability capping of clay or similar material, to limit further downward percolation into the contaminant. This is especially desirable in areas where the contaminant is underlain by permeable strata, unlike the industrial sites on alluvial clay where break-layers alone have proved effective (Section 19.1.1).

19.2 Excavation and disposal

19.2.1 Advantages and disadvantages

This is the second of the two reclamation methods in current use in the UK for contaminated sites, and involves the removal of contaminated material to a suitable disposal point and its replacement by clean imported material. It is the method approved for the majority of contaminants under the Building Regulations, 1985 (Table 24), unless expert advice is taken on alternative methods. Advice on the suitability of imported fill in given in Section 19.9.

Table 24 Possible contaminants and relevant actions (from Approved Document C1/2/3, DoE, 1985b)

Signs of possible contamination	Possible contaminant	Relevant action
Vegetation (absence, poor or unnatural growth)	Metals, metal compounds Organic compounds, gases	None Removal
Surface materials (unusual colours and contours may indicate wastes and residues)	Metals, metal compounds Oily and tarry wastes Asbestos (loose) Other fibres Organic compounds including phenols Potentially combustible material including coal and coke dust Refuse and waste	None Removal, filling or sealing Filling or sealing None Removal or filling Removal or inert filling Removal
Fumes and odours (may indicate organic chemicals at very low concentrations)	Flammable, explosive and asphyxiating gases including methane and carbon dioxide Corrosive liquids Faecal, animal and vegetable matter (biologically active)	Removal The construction is to be free from unventilated voids Removal,* filling or sealing Removal or filling
Drums and containers (whether full or empty)	Various	Removal* with all contaminated ground

*The local authority may require removal to be done by a specialist.

Thorburn and Buchanan (1983) report on a typical situation where no other expedient would suffice. Preliminary clearing of a 6-ha gas works site had left up to 6.5 m of randomly compacted, contaminated ground and spoil heaps, impregnated with tar products (Figure 38). Ground improvement was essential before building, but the tar impregnation ruled out covers as well as the use of vibro- or dynamic compaction. The only option was to remove unsuitable materials and replace with 2 m of granular materials.

Properly executed, excavation and disposal provides a permanent solution for an individual developable site, at the expense of creating an equivalent problem at the disposal point. It may thus be the ideal, though costly, solution for some types of site. On the other hand, several inherent drawbacks have tended to make it a less-favoured option in the UK than simple covering of the contaminant *in situ*.

19.2.2 Excavation

This is not so simple an operation as on a greenfield site and is troublesome where the water table is above the bottom of the excavation. Apart from the risks to site personnel and the continuing need for dust suppression, excavation invariably results in delays as unexpected buried features are encountered. Figure 12 illustrates the type of obstacles to excavation at a gas works at Gateshead. This site was eventually reclaimed by covering with a granular break-layer over the contamination (see Figure 35).

Figure 38 *Tar-covered bricks used as fill in an underground tank (photograph by courtesy of Environmental Safety Centre, Harwell Laboratory)*

More serious is the inevitable disturbance of the contaminants and exposure to air and water which can create higher contamination levels. The increased acidity and contamination levels in surface water drainage given below resulted from merely excavating trial pits on a 10-m grid (Cairney, 1985).

	pH	CN (mg/l)	phenol (mg/l)	SO$_4$ (mg/l)	As (mg/l)
Before	7.6	108	14	7600	12
After	2.3	357	64	3130	31

It is also possible for undetected pockets of contamination to be left in place, either within or beyond the confines of a proposed development, with the attendant risk that migrating groundwater may carry pollution into the clean imported material. Barrier walls or diaphragms may be needed to protect the cleaned areas and surface water drainage. Moreover, the mechanical stability of the material left in place should be preserved, and nearby buildings may need to be shored.

Excavation volumes may be large and difficult to assess in advance. In the London Borough of Barking and Dagenham a scheme to build 234 houses on 2.5 ha of an old tar distillery was preceded by reclamation involving the removal of 5000 m^3 of material containing identified contaminants (Armstrong, 1983). It later became evident that the reclamation was inadequate, and a volume of 90 000 m^3 was finally removed, requiring sheet-pile and bunded clay barrier walls to prevent ingress of groundwater to the contaminated ground. The operations had a temporary but unwelcome environmental impact from the diffusion of dust and fumes over the neighbourhood.

19.2.3 Disposal

The Control of Pollution Act 1974 prohibits the disposal of all 'special wastes' except on tips licensed to accept them. The number of such tips is limited, whereas the area of contaminated land available for reclamation is expected to increase considerably, with consequent severe competition for available tip space.

The tip operator requires to know in advance the quantity and composition of every load as well as the total volume to be delivered to the tip. This generates a considerable administrative problem for the developer or his contractor.

Many licensed tips cannot accept large volumes at a time. This may prove to be a complication in the development of large sites. On the other hand, a large site may provide more opportunity for disposing of its own contaminated material in other parts of the same site. For example, at Thamesmead (Lowe, 1984b) permission was granted to use excavated contaminated material behind the flood protection embankment. This special permission meant that the site was itself considered as a 'licensed' tip for Thamesmead's own waste. Generally, such permission may not be forthcoming, so that 'special waste' cannot be reburied on the site but must be transported to a tip licensed to receive it.

The large volumes to be disposed of to a tip can create serious traffic problems. An excavation of 1 m depth over 4 ha produces 40 000 m^3 spoil, equivalent to some 5000 journeys by 20-tonne lorries, with a similar number to import clean replacement material. The lack of suitable disposal tips in the immediate locality can result in long-distance haulage. In the case of Barking and Dagenham (Section 19.2.2), the local authority was obliged to limit the environmental impact by restricting traffic to 100 vehicle journeys per day.

19.3 *in-situ* treatment

19.3.1 Potential for application

Treatment *in situ* might appear an attractive alternative to removing or covering contamination and should provide a positive and permanent cure. It is free of the disadvantages of excavation and disposal and need not affect ground levels. It is now the subject of considerable research overseas (Sanning, 1985).

At present, however, *in-situ* treatments are expensive, difficult to apply, and of uncertain effectiveness. None of the available techniques appears to be generally applicable to large areas of land containing mixed industrial contaminants. These methods are more likely to be suitable for treating specific, identifiable contaminants of small extent (or homogenous contaminations of larger extent) as part of an overall treatment by other methods.

Well-established technologies, which have been applied to the *in-situ* treatment of contaminated land, include:

- Cement grouting to reduce permeability
- Chemical injection for neutralising specific contaminants
- Surface application (ploughing)
- Ground leaching.

19.3.2 Cement and chemical injection

Cement grouting. On contaminated sites cement grouting may have the following applications:

- Reduction of leachability of contaminant and/or containment of leachate
- Control of surface water ingress and groundwater movement
- Sealing around piles to prevent leakage to aquifer
- Protection of services, pipes, foundations against degradation in aggressive soils.

On fill sites it may have additional applications:

- Densification of fill
- Methane control
- Extinction of fires (uncertain effectiveness)
- Induction of anaerobic conditions.

Grouting and other forms of pressure injection are difficult to perform at depths with less than 2-m cover, that is, in the zone in which the majority of contamination of interest to building sites resides. Temporary surcharging of the ground may be a feasible solution.

Jet grouting (Huck *et al.*, 1980; Childs, 1985) has been proposed to form an underseal to buried contamination to protect aquifers, but it is difficult and expensive and of uncertain efficacy. It is also suitable, and more effective, for the creation of vertical barriers to regulate the lateral migration of groundwater. See Section 19.8.2.

Chemical injection has the potential for:

- Neutralisation of contaminants
- Introduction of microbiological agents
- Aeration (for microbial action).

Unfortunately, it is virtually impossible to identify a specific contaminant and to bring the desired quantity of injected chemical into intimate contact. Reactions may take a long time due to the slow movement of groundwater, and the reaction products themselves may be harmful.

19.3.3 Surface application

Deep ploughing (to 1.25 m), with or without incorporation of lime or other neutralising agent, may be useful in inverting the top layers, with resulting dispersion/dilution or neutralisation of surface contamination. It has been used experimentally to treat surface concentrations of heavy metals in land covered with sewage sludge (Wood and Ross, 1979). For the liming of soils contaminated by metals, regular dosage is required.

Where there is continuing generation of acid in the ground during, for example, the progressive oxidation of pyrites (Figure 11) in exposed coal-mining spoil, the pH of lime-treated ground tends to revert. Lime treatment thus requires periodic renewal, beyond that necessitated by natural lime loss in all soils.

19.3.4 Ground leaching

This promising technique is already used to clear salinity from land subject to marine flooding. It was used over a period of five years in Sweden (Lindfors, 1979) to reclaim the contaminated site of a former pesticide factory. Water percolated through the ground from a trench along the middle of the site. The resulting leachate was collected in deep drains and treated with activated carbon, and was then discharged to a sewage works for BOD reduction.

19.4 Soil processing: on- or off-site

19.4.1 The process and its potential

Soil treatment is intended either to remove contaminants completely (the ideal solution) or to reduce their availability. If this cannot be accomplished *in situ*, the soil must be excavated for processing, either on- or off-site. Processing on-site, if practicable, would avoid the severe problem of off-site traffic. Experience in the Netherlands and West

Germany, however, shows that local environmental considerations, and local opposition generally, will normally enforce treatment at off-site facilities. This has the advantage that such facilities can receive material from a variety of sites.

A variant of the process is to excavate contaminated soil and to mix it with clean soil, either imported or from elsewhere on the site. The soil, diluted to an acceptable concentration, is then distributed on-site and covered with clean material, if necessary.

19.4.2 Practical applications

Ravenfield tip. A notoriously toxic site of 2.6 ha was the result of uncontrolled tipping of acid tars (32% H_2SO_4), oil-soaked Fuller's Earth, industrial wastes and slags from Rotherham (Khan, 1979). About 10 000 tonnes of heavily contaminated fill were treated to reduce the availability of contaminants. The waste was excavated in batches and mixed with pfa and an excess of lime. The neutralised material, friable and innocuous, was then replaced in layers. In this somewhat special case, neither the neutralisation products nor the excess lime constitute a noxious hazard for the proposed development (landscaping without buildings).

Greenbank Gasworks. In 1985, there was a contract for the microbiological on-site treatment of the 10-ha Greenbank Gasworks at Blackburn. Treatment involved the biodegradation of the organic gasworks residues by the extant microbes, which were already consuming the toxic compounds although at a slow rate. Laboratory-cultured and intensified bacteria were to be injected together with nutrients into the toxic wastes, already excavated and formed into stockpiles on site.

19.4.3 Future development of processes

Rulkens *et al.* (1984), de Leer (1985), Assink (1985), de Kreuk (1985) and Hilberts *et al.* (1985) give details of processes currently under research or experimental operation in Europe, including:

- Extraction leaching of metals as soluble salts
- Mechanical separation by sedimentation
- Flotation—aqueous suspension frothed by air injection
- Thermal treatments:
  —low-temperature evaporation of volatile organics
  —high-temperature destruction of contaminants
- Steam stripping to volatilise organic components
- Chemical/redox reactions to yield harmless products
- Microbiological methods:
  —anaerobic organic breakdown to CO_2, CH_4, H_2O, N_2
  —aerobic formation of leachable salts (such as sulphides into sulphates)
  —co-disposal of domestic/industrial wastes
- Stabilisation to immobilise contaminants for safety.

19.5 Control of landfill gas movement

This section is addressed in particular to the development of those earlier wastefills deposited without special care prior to about 1975. Before building development is contemplated on any gas-hazardous site, careful note should be taken of the reservations expressed in Section 17.8 and elsewhere. In cases where building development is judged to be permissible, the safeguarding of a building can be effected in two distinct ways:

1. Control of gas movement in the ground
2. Safe design of the building itself (dealt with separately in Section 20.3).

These are not alternatives. Either safe design will be sufficient (in less severe cases of gas emission) or gas movement will have to be controlled (in more severe cases) prior to developing the building in a safe manner.

19.5.1 Basic concepts

Gas movement control may be required to reduce the quantity of and/or to impede access of underground gas to buildings (existing or proposed). Typically, it makes use of two proven technologies:

1. *Passive*: gas drainage through venting trenches, with or without associated vertical cut-off barriers. This is particularly important as a peripheral protection, completely surrounding a hazardous area or, more rarely, the buildings themselves
2. *Active*: gas extraction by pumping, with or without flaring or gas collection. This is almost always associated with peripheral vertical cut-off barriers.

In this context a distinction must be made between modern landfills deposited since about 1975 and wastefills of earlier date.

Modern landfills. These fills, deposited in response to the Control of Pollution Act 1974, are constructed with special care to facilitate eventual reclamation and redevelopment. The active and passive systems are generally integrated and may include specially constructed pumphouses for gas extraction from a network of perforated pipes, laid at successive levels in gravel-filled trenches as landfilling proceeds. Gas-drainage wells may be included at nodes of the pipe network, and there may be facilities for collecting or flaring the gas.

An important peripheral protection is the incorporation of a venting/cut-off lining around the boundary of the filled pit or around the gas-generating zone. It is intended to form a preferred path for intercepting and guiding migrating gas, so as to prevent its entry into sensitive areas beyond the confines of the gassing zone. It consists of a wall, about 1 m thick, of gravel or coarse broken stone against the sides of the pit, built up in stages up to final ground level as filling progresses. It should normally incorporate a low-permeability skin (e.g. a membrane or bentonite or clay lining) on the side away from the source of gas.

Waste Management Papers 26 (DoE, 1986) and 27 (HMIP, 1989) deal in some detail with recommended procedures for the formation and management of modern landfills. Here, gas movement control is in the province of waste management, and is unlikely to be of interest to a building developer. The engineer should, however, ensure that site gas management is satisfactory before recommending building development on or adjacent to such sites.

Earlier wastefill sites. Very few wastefills of earlier date, and now completed and abandoned, have the above protection systems. Nor is effective gas management likely. Decisions on whether to build on or adjacent to such gas-hazardous land must consider the extent to which it may be necessary to install protective measures and undertake gas management.

Sections 19.5.2 and 19.5.3 describe the works needed for the later installation of passive and active protection at an existing wastefill.

19.5.2 Passive protection: peripheral venting/cut-off trenches

A passive peripheral protective system may be required wherever building is contemplated within the zone of influence of an actively gas-generating source. If it does not already exist as a pre-installed lining wall, progressively constructed in advance of the rising fill, as above, it may have to be installed before building work, in specially constructed trenches.

Venting/cut-off trenches, filled with coarse gravel or broken rock, have been used as preferred paths for intercepting and guiding migrating gas, and have proved useful in suitable conditions for the protection of property in gas-hazardous areas. Trenches can be conveniently the width of a digger bucket, or about 1 m. Where the adjacent land is permeable to gas, the system should incorporate a plastic membrane, or bentonite or clay barrier, on the side away from the source of gas.

Depending on site conditions and local authority requirements, the granular filling may be left exposed at ground level, or may be capped with clay or other low-permeability material.

If left uncapped, the trench will require frequent maintenance at ground level. This involves stripping off the top 100–200 mm of gravel, cleaning and replacing.

If, instead, the trenches are capped, they will need venting by large-diameter wells or boreholes at frequent intervals; narrow-bore vertical pipes are not sufficient. London Scientific Services have recently specified a useful variant on the venting system, incorporating a precast 'manhole' with perforated sides, of diameter equal to the width of the trench and with a protected vent safely above ground level (Carpenter and Lowe, 1988).

Trenches are difficult to install in wastefill and are usually constructed in the natural ground as close as possible outside the fill boundary. Excavation to install the permeable gas-venting material is only practicable to depths of about 8 m (HMIP, 1989). Below that depth a grout curtain may be needed (Barry, 1987). Venting trenches are a useful and often essential expedient but they are not an infallible remedy, and each case must be carefully considered. Waste Management Paper 27 (HMIP, 1989) illustrates the principles of venting/cut-off trenches. Their installation, however, should be entrusted to those with expertise in this field.

19.5.3 Active protection: pumped gas extraction

In cases where site investigation has revealed gas generation of an intensity too great to be dealt with by passive venting, as above, recourse to pumped extraction may be necessary. This may be required for safeguarding a building or may be demanded by the statutory authority for nuisance abatement. Its reliability in performance must therefore be assured.

The execution of a pumped extraction scheme calls for the collaboration of a specialist contractor. The work involves first a pump/flare exercise intended to demonstrate the quantity of gas likely to be generated in the future, and the probable intensity of surface emission. If the tests reveal the necessity, the next stage is the installation by the specialist contractor of an engineered system for gas extraction, including, as necessary, gravel-filled trenches, cut-off barriers, perforated pipes, boreholes, pumping station with extraction fans and, perhaps, flare stacks.

Pumped gas extraction is costly to install and operate. A building development which requires such a measure for its viability may prove economically unattractive.

19.6 Treatment of potential fire hazards

19.6.1 Options

If the appraisal leads the engineer to suspect a hazard from combustible materials in the ground two options only are open to the developer:

1. Treat the ground to eliminate the hazard
2. Abandon the proposed building development.

To contemplate building directly over a potential combustion hazard without treatment is an unacceptable risk, and would not receive local authority consent.

19.6.2 Precautions

In all works carried out on fire-hazardous sites special care must be taken not to admit air to spontaneously ignitable materials or in any other way to provoke ignition by careless use of fires or other sources of heat (Section 12.9). To extinguish underground smouldering, once started, may prove impossible and will certainly be more onerous than preventing ignition. All trenches opened on the surface should be backfilled the same day or as soon thereafter as possible. Cable tool boreholes have not been known to start fires, but open-hole rotary drilling with air-flush can do so.

Whenever combustible material is to be left in place the relevant facts should be recorded on the property deeds as a warning to future users. Where possible, the layout plan should preclude the siting of boiler furnaces or other heat sources above affected areas.

19.6.3 Category A: Spontaneously ignitable material

The most effective treatments of the ground under a proposed building are:

- Removal of all combustible material or
- Mixing and dilution with inert materials so that the combustible material cannot ignite or, if ignited, cannot sustain combustion.

Both these options may prove unacceptably costly.

Under favourable conditions it may be cheaper to leave spontaneously ignitable materials in place, and to compact the upper layers to reduce the air supply to below ignition requirements. There remains the danger that a future disturbance may admit air and unwittingly trigger off ignition. Furthermore, compaction alone may not be sufficient to prevent ignition by an external source of heat or by a migrating underground fire.

Compaction is a tenable option only as long as any future excavation, or other disturbance, can be confined to layers safely above the compacted combustible. Any cover layer (Section 19.1) must be sufficiently thick to prevent overheating of the combustible material by an external source. A net unpenetrated thickness of 1 m of material (granular or cohesive) is generally found to be adequate. Granular material will not be effective by itself in restricting the air supply, and should be supplemented by a 100-mm layer of damp clay or similar fine-grained material. Until adequate covers are installed, surface fires must be forbidden. 'Fire fences' (see Section 19.6.5) must be incorporated wherever migrating underground fire is a potential hazard.

19.6.4 Category B: Non-spontaneously ignitable material

Treatment is generally as for Category A material except that removal or dilution, although advantageous, may not always be obligatory. The essential is to deny access of air and/or heat to the combustibles. Compaction and/or covering may be sufficient safeguard as long as fire fences are provided against migrating underground fires. With a high water table treatment may not be necessary. It will, however, be essential to ensure that the material will never dry out during the whole life of the project.

19.6.5 Underground fire fences

Where potentially combustible material has been left in place under a building, neither compaction nor covering is proof against migration of underground fire. A surrounding cut-off wall, termed a fire fence, may be required to protect the target.

A fire fence is usually formed of inert, impermeable fill in a trench taken down to a level below any fire hazard. In shallow fill the trench may advantageously be taken down to original ground or to the lowest probable water table. Material and thickness should be such as to ensure an adequate temperature gradient so that the material remains always below its ignition point. About 1 m of low-permeability material (clay, pfa or bentonite slurry) is considered adequate.

Weatherley (1979) reports the use of continuous pfa piers under bungalow strip footings, which also functioned as a fire fence, permitting some of the combustible site material to be left *in situ* under the buildings (see also Section 18.3).

19.6.6 Dealing with underground burning

Underground burning, or even smouldering, is extremely difficult to control, and no routine method at present available is dependably effective. Restricting the air supply may prevent ignition but is unlikely to stop combustion once started.

Water should never be used to extinguish underground burning without expert advice, due to the risk of increasing combustion through washing out new air-access channels. Furthermore, in the case of incandescent underground fire, the heat output may be high enough for the production of flammable 'water gas', which may migrate and ignite elsewhere on the site.

It has proved possible on occasions to excavate burning colliery spoil with conventional earth-moving plant. The spoil is allowed to cool and is then re-compacted once it has burnt out. The feasibility of this procedure depends on the site layout, wind direction, and behaviour and temperature of the spoil.

Grout may be injected into burning (or potentially combustible) fills, but success is not a foregone conclusion. A temperature survey is carried out on a grid-pattern to define the combustion zone, around which a primary ring of grout holes is drilled. Cement grout is injected under pressure in an attempt to form a cut-off or to fill the voids, thus reducing permeability. Limestone dust is a useful additive as its decomposition is endothermic (cooling effect) and produces carbon dioxide, which may help to deny air to the burning zone. Frequent temperature measurements are necessary to monitor the effectiveness of the treatment as the grouting proceeds. Secondary rings of holes are then drilled, closing in successively towards the hottest part of the zone.

Grouting of shallow combustible material suffers from lack of cover (Section 19.3.2). Surcharging is a useful expedient, and has the added advantage of safely compacting the combustible material. On the other hand, the grout holes must be drilled through the full extra thickness of the surcharge, and may be lost or damaged if the surcharge moves or settles as a result of the underground combustion.

19.7 Combating aggressive ground conditions

19.7.1 Basic concepts

The level of the water table has an important influence on the potential for aggressive attack. In dry conditions water-soluble chemicals are considerably less aggressive. Care must be taken to ascertain whether an apparently low water table is the result of a temporary condition such as local pumping, and whether capillary suction may induce contaminant migration.

Excavation and complete removal of a contaminant would be the most positive way of preventing contact of the chemical with surface and sub-surface targets and the aquifer, but complete removal cannot be demonstrated with certainty. Instead, provided the water table can be held down, it may be sufficient to bar the chemicals from upward migration by partial removal of the overlying ground and replacement with clean material to a sufficient depth (perhaps 0.5–1 m). Deep drains may be used to depress the water table and/or protect against lateral migration, but may be too deep to discharge without pumping. Surface slopes should be graded, compacted and/or treated to reduce ponding and percolation into the ground. Lateral migration of groundwater may be impeded by a membrane or cut-off wall of bentonite or other material of low permeability, but the effectiveness of this measure may be reduced by excavations for pipes, etc. (see also Section 19.8.3).

In an ideal situation where the protection of foundations, services, etc. were the only consideration and percolation always downward, a simple granular infill would be the best replacement material. However, other conflicting requirements (such as protection of the aquifer) may intervene, and the basic principles of multi-purpose cover layers (Section 19.1) must be carefully considered. A clear understanding of the hydrogeology of the site is necessary before definitive decisions can be made.

19.7.2 Practical realisation: protection of concrete components

Where concentrations of aggressive contaminants, sulphates, etc., or pH values lie outside the generally accepted values for concrete (Section 20.4), protective measures will be required for foundations, services, manholes and similar concrete targets. Proper protection will need some or all of the following:

- Removal of contaminant from direct contact with concrete
- Location of concrete above highest water table
- Location of concrete above migration level of aggressive contaminant
- Protective coatings (considered as a design measure).

Removal of contaminant. Removed material should be replaced by clean granular material, with under-drains if levels allow. (See Section 19.9 for a description of the suitability of materials as imported fill.) Complete removal is almost impossible to demonstrate, and reliance should not be placed on removal alone; other protective measures should also be incorporated.

Location above water table. Deep drains may assist in maintaining a permanently lowered water table. Control of water table and groundwater migration by pumping and barriers (Section 19.8.3) may be economically feasible on large sites (as Stockley Park case history—Section 19.8.3).

Location above contaminant migration level. In cohesive ground liable to soil suction a break-layer (Section 19.1) may be required, of granular material usually 0·5–1 m thick above the highest water table. It is unlikely to be needed, even in cohesive ground, if the target is at least 4 m above the highest water table.

Protective coatings. Whatever other measures are adopted, applied surface protective coatings are an advisable additional precaution. These are dealt with in Section 20.4.

19.7.3 Services

Until recently, services in aggressive ground were laid in oversize trenches and backfilled with clean, inert material. Water authorities are now less likely to accept this practice (Section 13.2.3), and may demand special measures, such as a sealed conduit to contain all services or the installation of services within a clean designed cover (Section 19.1).

19.7.4 Electrolytic action

Metal pipes, sheet piles, etc. may be locally corroded by electrolytic reaction with dissimilar metals present in the ground, either as scrap or otherwise. Bimetallic action should be avoided by excavation of suspect ground, provision of protective coatings or appropriate design detailing. Backfill, especially clay, around metal products should be uniformly compacted, as uneven compaction may facilitate electrolytic action.

19.8 Protecting water resources, foundations, etc. against leachate

Section 17.9.1 explains that water authorities are likely to demand measures to treat leachate from a developing derelict site or to reduce its volume, to protect local water resources and the aquifer. Some of the procedures outlined below are also suitable for the protection of foundations and other underground components from aggressive groundwater.

19.8.1 Leachate treatment and reduction

Treatment processes for landfill leachate, involving aerobic/anaerobic recycling through the fill, with pH control and sewage sludge seeding, active carbon filtration and ion exchange are proven methods for leachate improvement (Jarrett, 1979; Barber, 1983; Cheyney, 1984). They belong, however, to the disciplines of waste management and are unlikely to concern a building developer.

Protective measures which could be required of a developer involve restriction of leachate production to an acceptable level and control of leachate migration. Simple expedients toward that end include covers, water-shedding, peripheral drains and surface vegetation. These are unlikely to be sufficient alone, and some of the more sophisticated measures outlined in Section 19.8.3 will probably be called for.

19.8.2 Static approach: encapsulation of contaminant in situ

Where a contaminant is not to be removed, its encapsulation to protect the local aquifer might appear to be a desirable option. However, with the limited technology at present available, reliance on encapsulation *alone* in perpetuity is an unrealisable and dangerous ideal.

Encapsulation of an extensive contaminant would demand permanent integrity in each of the following components, designed to prevent any contact between contaminant and migrating groundwater:

- Horizontal covers to exclude surface infiltration
- Vertical barrier walls to exclude migrating groundwater
- Bottom seal (floor) under contaminant.

Covers can be effective for a variety of designed purposes, and are generally used in the UK for purposes not relying on encapsulation. Their function and design are dealt with separately in some detail in Section 19.1.

Vertical barriers are susceptible to ground movements and contaminant attack and their effectiveness may range from good to only fair. They are more useful in regulating mass movement of groundwater/leachate rather than complete encapsulation. They can be formed of slurry-filled trenches, grout curtains, piled walls and membranes, installed by proven technologies.

Bottom sealing is very difficult to execute, and its effectiveness cannot be demonstrated. The least unreliable method is to seal vertical barriers into underlying impermeable strata, where possible. The alternative, of forming a 'floor' under the contaminant by newly developing techniques of horizontal grouting or by jetting ('kerfing') and slurry filling, is at present fraught with uncertainty.

19.8.3 Dynamic approach: control of groundwater regime

Basic concept. Contemporary practice, particularly in continental Europe and America, regards local groundwater, contaminant, leachate-plume, aquifer and surface water resources as parts of a dynamic system, amenable to some degree of imposed hydrological control. Incorporation of covers, vertical barriers and hydraulic measures into the system imparts some local control over water tables, rates and directions of groundwater flow, etc. Complete impermeability is not expected of the barriers. Instead, they should be adequate to prevent mass transfer of groundwater or leachate, and thus to reduce infiltration and migration rates to acceptable values.

The control of a groundwater regime is a very complex and difficult procedure which requires considerable experience, detailed knowledge of local hydrogeology, expert design (generally by mathematical modelling, see Section 17.9.2), as well as extensive and enduring engineering intervention and monitoring.

Practical realisation. Childs (1985) gives a wide-ranging and instructive review of the installation and operation of modern systems to safeguard aquifers and surface waters. The following notes are based on Childs, but his original paper should be studied whenever leachate control is under serious consideration.

A protective hydraulic system may include static covers and barriers, monitoring and pumping wells, and provision for treatment of pumped leachate. Childs discusses pumping techniques and barriers for inducing local changes in water tables, leachate plumes and groundwater flow patterns. Depending on requirements, changes in regime may be exploited for, *inter alia*:

- Limiting, or preventing, contact between migrating groundwater and contaminant
- Protecting surface environment by drawdown of leachate plume so as to prevent its entry into lakes, streams, etc.
- Protecting aquifer by producing localised upward hydraulic gradients to detach leachate plume from main body of aquifer
- Containment of leachate plume by pumping from a ring of wells round contaminant, to protect drinking wells, streams, and other sensitive targets, and slowly deplete the contaminant source
- Increasing travel distance through soil and reducing leachate concentration, to assist attenuation by natural processes
- Slow leaching away of contaminant at a controlled and safe rate along planned path to permit its extraction or natural dispersion downstream
- Lowering contaminated water table to prevent contact with sensitive foundations, services, etc. (see Section 19.7.2).

The pumped leachate may be discharged to a foul sewer or approved water course or, after appropriate treatment, to ground or to aquifer recharge. Monitoring is required to maintain a proper regime, and pumping rates will vary accordingly. Monitoring also serves to sound an advance warning of changes or danger, so as to permit the timely activation of countermeasures (such as increased pumping upon deterioration of a static protective barrier system).

There may be advantage in the slow leaching away of a contaminant if this results in the eventual exhaustion of the source (see also Section 19.3.4). This is a very different concept from the indiscriminate leakage of contaminant to the detriment of other areas. The rate of leaching and the path to a collecting or dispersal point must be carefully controlled, and leaching must not lead to the accumulation of contaminated water at some other sensitive point inaccessible to pumping. Where, alternatively, reliance is placed on attenuation through the cleaning action of the natural ground, leaching must be so slow as not to exceed the capacity of the ground.

Case history. The Stockley Park, Hillingdon, project has involved the redevelopment, under private financing, of a 150-ha quarry site, heavily contaminated to a depth of 15 m with industrial, commercial and putrescent domestic wastefill. Reclamation has necessitated 4 million m^3 of earth-moving. For the protection of a new 30-ha factory area, migrating leachate is intercepted by a deep drain from which it will be pumped away for 20 years or so, until monitoring has shown biodegradation and contaminant depletion to be acceptably complete.

19.9 Imported fill materials

Import of clean materials may be required for reclamation purposes. Advice on the suitability of materials as fill is contained in the DTp specification for Highway Works (1986). In decisions on the importation of fill materials, the recommendations of BRE Digests 274, 275 and 276 (BRE, 1983a–c) should be carefully considered. Sherwood (1987) gives valuable information on all commonly available wastes for fill.

19.9.1 Imported fill

Provided that sufficient thickness of topsoil/subsoil is specified, a wide range of materials is suitable for the bulk infill. The material should be spread and compacted as in good engineering practice. A granular material is preferred, with a clay content not greater than 10%. All materials should be checked for contamination at source, not after arrival.

The suitability of various materials as imported fill is given in Table 25. The materials marked 'with caution' may be used for lower layers of a fill but with due regard for the specific defects noted.

Table 25 Suitability of materials as imported fill

Material	Suitable	Remarks
Natural soils	Yes	As in Specification for Highway Works (DTp, 1986)
Quarry waste	Yes	
Building or demolition rubble	Yes	If free from gypsum, plaster or lead
Blast-furnace slag of recent origin	Yes	
Pfa from bituminous coal	Yes	
Unburnt colliery shale	With caution	Check combustion potential, sulphates
Old blast-furnace slag	With caution	Check chemical stability
Incinerator ash	With caution	Check heavy metal content, sulphides
Pfa from brown coal	With caution	Check free lime content
Rubble containing plaster	With caution	Check sulphate potential
Burnt colliery shale	With caution	Check sulphate potential
Steel-making slags	No	Chemically unstable
Non-ferrous slags	No	Toxic
Dredged silts from waterways	No	Heavy contamination and physically unsuitable
Domestic waste	No	
Biodegradable materials	No	
Combustible materials	No	
Pyritic shales	No	
Materials containing soluble sulphates (Na, K, Mg)	No	

If the size of the site is appreciable, and subject to the approval of the waste-disposal authority, hollows and pits may be partially filled with contaminated material from other parts of the site, diluted to an acceptable concentration or buried at sufficient depth.

19.9.2 Imported topsoil

Experience shows that gardens and allotments typically require only 60% of the topsoil available on a greenfield site. Clean topsoil arising from road and building works, for example, may therefore be sufficient to replace that lost in the removal of contamination. Topsoil should not be compacted.

Topsoil local to the site may be contaminated and of poor quality while good topsoil is becoming increasingly scarce and expensive to import. As an economic alternative, clean local subsoil may be converted to a good growing medium in two or three years by adding nutrients, seeding with ley grass and grazing (Keeble, 1983). Imported topsoil should always be checked (at source, not after arrival) for conformity with BS3882: 1965.

Sewage sludge may be used as a constituent of topsoil, provided due allowance is made for any heavy-metal and other toxic content. It has been successfully used in the UK as a source of nutrients and organic materials on mineral wastes. When spread in thick layers, however, it may create anaerobic conditions and inhibit plant growth.

19.9.3 Recycled industrial waste as fill material

Commercial processes exist today for supplying inert slurry for infill on land-reclamation sites. The raw material is industrial waste, treated at a central processing plant and incorporated into a cementitious mix. The end product, sold under trade names, is said to set to a firm solid in 2–5 days.

19.9.4 Pulverised fuel ash as fill material and topsoil

Sources of suitable pfa. The following notes refer only to pfa produced by UK power stations burning bituminous coal. They do not apply to ash from brown coal or to wastes from FGD (flue-gas-desulphurisation) plants, due to enter operation in the mid-1990s. The latter will include environmentally sensitive sludges requiring specialist treatment.

UK power stations produce about 12 million tonnes a year of ash, including pfa (8–10 million tonnes), furnace bottom ash and clinker, of which the market gainfully absorbs about 40% for manufacture of cement and concrete products and 10% for reclamation landfill. The surplus 6 million tonnes a year must be disposed of to land, usually in unsightly and environmentally damaging lagoons and tips near the power stations.

Bulk infill. In cases where its properties are suitable there should be great environmental advantage in using surplus pfa for reclamation landfills, thereby avoiding the need to quarry fresh material, and eliminating unsightly deposits near the stations. Since suitable fill sites are often at large distances from a power station, careful transport management is required to reduce the level of general nuisance from dust, etc. Pfa may be pumped as a slurry (60% water) for distances up to 20 km or it may be delivered by rail or road, conditioned to approximately 20% moisture content.

For general bulk infill pfa is a valuable material. In addition, its pozzolanic setting properties find several special applications, and its low bulk density is advantageous for embankments on poor ground. Some caution is, however, needed in its use, with attention to its chemical and physical nature.

Generally, pfa is neutral-to-alkaline and contains significant amounts of trace elements (As, Bo, Cd, Pb, Se and V). It also exhibits some radioactivity, but usually at levels within the range reported for natural soils (see also below on radon emission). In the dry state it is also a very dusty material, and is often supplied conditioned. For bulk infill it should be buried at least 0.5 m below any exposed soil surface. A granular break-layer may also be needed against upward migration, whether of contaminant existing in the underlying formations or imported with the pfa itself. There is also some evidence for caution in using it as bulk infill in residential developments, even where it is overlain with topsoil (Smith,

1987). Protection of ground and surface waters from saline leachate generated within the pfa requires good compaction of the ash, and sufficient cover and surface slope to encourage run-off instead of infiltration.

Topsoil substitute. Pfa has been proposed as a topsoil substitute. It should, however, be used with caution on account of its variable trace metal content, and permission should be sought from the environmental health officer and/or the water authority. Chemical analysis will be required for the trace metals listed above, all of which can be harmful to plants. Of these elements, vanadium is not yet listed in the ICRCL trigger values (Section 17.4.1), and boron is said to reduce by weathering to acceptable values in about two years. Pfa is also sterile and devoid of nitrogen. There is also some evidence that mixing ash with topsoils increases the availability of some elements (Smith, 1987).

Physically also pfa is less than ideal as a topsoil. Of very fine grain, it has a high soil-suction characteristic liable to induce upward migration of contaminants. A break-layer is possible under the pfa, but would be ineffective against contaminants within the pfa itself. It would also require a filter against loss of fines, at sufficient depth to prevent its disturbance by agricultural operations.

Safety of site workers. Although trace metals may require measures for safeguarding sensitive end uses from long-term hazards, they are not considered to be sufficient to put temporary site workers at risk for short-term exposures. In case of longer-term exposure of site workers, however, the HSE may demand protective measures (water spraying, etc.) against the discomfort of inhaling pfa dust. The alkalinity of pfa may present a hazard to workers through skin contact, especially those already suffering from dermatitis induced, for example, by chromium in Portland cements (Smith, 1987).

Care is needed in excavation and other work on pfa dumped in lagoons, where apparently firm ground may be merely a crust formed over underlying liquid material.

Radon emission from pfa. There is evidence that pfa deposited in the ground emits radon gas (National Radiological Protection Board, 1986). The conclusions of NRPB are generally that there is no danger to the public out of doors, although there may be potential increase in exposure of occupants in dwellings over ash-disposal sites from increased radiation flux out of the ground. This is not of great radiological significance, and no action is considered necessary in existing dwellings. NRPB further conclude that there is no significant radiological hazard to persons working at newly restored or working pfa-disposal sites.

For proposed new buildings simple precautions may usefully be considered at the planning stage to keep radiation levels (from accumulating radon gas) as low as reasonably achievable. Precautions against radon infiltration should be similar to measures described for landfill gas: underfloor voids with ventilation, crack-free concrete floors with sealing membranes, etc. (Section 20.3).

20. Design measures

20.1 Design against settlement

20.1.1 Basic concepts

Defensive design for protecting buildings on fill accepts that some residual settlement is inevitable, even after remedial treatment, unless piling is adopted. Foundations bearing on fill should be designed to permit settlement without subjecting the superstructure to damaging differential movements or unacceptable tilt.

Depending on the depth of fill, and the extent, height and sensitivity of the structures, there are three choices for the design of foundations (Table 26):

1. Positive support: piling
2. Floating support: rafts
3. Floating support: flexible with individual footings.

Framed structures. The tolerance of buildings to distortion has been investigated by several workers, and the limiting values have been discussed by Padfield and Sharrock (1983). In very simplified terms, the practical effect of these observations is that most normal framed buildings (of the type likely to be founded directly on fill) can tolerate a differential settlement of about 20 mm between columns. This sets the limits for flexible floating supports with individual footings. If this acceptable settlement is likely to be exceeded, good practice will require at least a raft foundation for low-rise structures of modest extent or piling for higher-rise or extensive structures.

Table 26 Preliminary guide to choice of foundation type on fill

Foundation		Fill		Normal applications
Concept	System	Condition	Depth[1]	
Positive support	Piling, piers, strip footings, taken down to natural ground	Immaterial	Shallow	2, 3
			Medium	Housing[3] High-rise structures Sensitive structures Process plant
			Deep	Housing High-rise structures Sensitive structures Process plant
Floating support (rigid)	Rafts[4]	Grade I	Shallow	Housing[5] Low-rise structures
			Medium	Housing[5] Low-rise structures
			Deep	Housing[6] Low-rise structures
Floating support (flexible)	Reinforced pad or strip footings (with non-suspended floors)[8]	Grade I	Shallow[3]	Housing[7] Low-rise, wide-span structural frames Light industrial units
			Medium[3]	Low-rise, wide-span structural frames Light industrial units
			Deep	Not appropriate

Notes: 1. *Depth of fill:* Shallow: 0–4 m; medium: 4–15 m; deep: over 15 m.
    2. Not normally economic.
    3. At around 3–4 m depth, piling (or pier and beam foundation) can be more economic than trench-fill or strip foundations (see Section 20.1.2).
    4. On shrinkable clays, rafts may be less economic than piling (see Section 20.1.2).
    5. Semi-rafts (as Section 20.1.3) also used.
    6. Old fills only, with self-weight settlement acceptably complete.
    7. Only with full-depth treatment.
    8. NHBC do not permit non-suspended floors for housing on fill (see Section 20.4.2).

This table is a preliminary guide only, and is not intended to be comprehensive. Final choice of foundation type will depend on site and treatment, codes of practice, etc.

Brick housing and other brittle structures. Brittle structures are considerably less tolerant of different settlements. Floating support on individual or reinforced strip footings can only be considered for buildings of modest extent on shallow fill which has received full-depth remedial treatment. On medium-depth or deep fills safe design will require a raft for houses of limited extent, or piling for more sensitive, higher-rise buildings.

Additional information. For dwelling houses and low-rise buildings, Tomlinson *et al.* (1978) give valuable information on foundations on various types of ground, including shrinkable clays, fills and subsidence areas.

20.1.2 Positive support: piling

Successful piling in fill presupposes full penetration down to underlying natural ground of adequate bearing value at a feasible depth. Provided piles are designed against all adverse features of a fill site, and are founded in natural strata, they are the most certain defence against settlement of the structure.

According to NHBC, small house builders may today find piling an attractive proposition in fills between 4 and 15 m in depth, as well as over 15 m, in part due to the recent considerable rise in land values. A factor also favouring piling, or piers, against trench fill or deep strip footings in fill around 3–4 m depth is the difficulty and cost of side support to a trench excavation while concrete is being poured. In addition, in shrinkable clay fills, the potential heave due to the removal of vegetation or settlement due to tree planting may render a raft solution less attractive than piling.

While piling will obviate settlement of the structure, problems may arise from settlement of the fill outside the building area. Service connections and the discontinuity of level at the building periphery are particular problems. Furthermore, piling may create migration paths for gas and leachate in landfill or allow air access to combustible materials. Seal-grouting may then be a troublesome necessity. On gassing sites, other design expedients will also be necessary (see Section 20.3). The piles themselves may require protection from corrosive attack (see Section 20.4). Piling may be impeded by old foundations commonly to be found in industrial dereliction or by blocks of concrete, stone, brickwork and similar obstructions in fill.

Piles should be designed in accordance with accepted methods but no positive frictional component should be assumed for the fill material. Instead, the design must allow for the 'downdrag' exerted on the piles by the settling fill. Conventional methods may be used in computing the downdrag, but the resulting design often leads to piles notably more robust than would be necessary for carrying the structural loads alone.

The use of slipcoats should always be considered as an alternative. Depending on their design, these can reduce downdrag effects by as much as 90% but do not eliminate them entirely. In practice, the formulation of a slipcoat is a compromise between competing requirements: reduction of downdrag; resistance to abrasion during driving; and, in certain cases, protection of pile from aggressive attack. Slipcoats can be applied as a coating to driven piles only. For bored piles a protective sleeve may be a difficult and costly expedient.

The choice of pile type is dictated by economics once the practical alternatives for the site have been identified. There is, however, no simple technical solution to the choice of piling systems, and each case has to be very carefully considered.

Obstructions will be a problem for all piling systems. Small obstructions may be displaced by driven piles, but the sort of obstructions likely to be encountered on derelict land will generally need to be excavated and broken out for any type of piling. In some cases pre-boring for driven piles may be necessary to ensure full penetration and avoid obstructions. There may be occasions when small obstructions can be more easily penetrated by bored piling, but at the cost of expensive and time-consuming techniques. Bored piles inherently provide the opportunity to establish with certainty that the fill has been fully penetrated.

On contaminated sites with aggressive ground conditions there may be advantages in the use of driven precast segmental piles. However, consideration must be given to the material through which the piles are to be driven and its effect on the integrity of any protective coating.

Incomplete penetration of the fill may invalidate any benefit from the adoption of piling. Smyth-Osbourne and Mizon (1984) report severe differential settlement of a single-storey factory on piles in an opencast fill of average depth 10 m deposited 24 years previously. Because of an erroneous site investigation, several piles stopped short of good bearing in original ground. Investigation showed that the settlement of the structure resulted from collapse settlement of the backfill, provoked by a rise in the water table upon cessation of colliery pumping in the area.

With properly designed and executed piling in fill, the design of pile caps, ground beams and floor slabs follows normal engineering practice (but see Section 20.3 for detailing on gassing sites).

20.1.3 Floating support: rafts

Limiting distortions. Most buildings can tolerate about 20 mm of differential settlement between columns. Furthermore, on uniform ground differential movements will normally not exceed 75% of the overall settlement (Section 17.3.9). On a greenfield site these two facts lead to the generally accepted safe limit of 25 mm total settlement for buildings on isolated footings.

However, fill sites are more heterogeneous and more liable to serious differential settlements than uniform compacted natural soil (Figure 39). Furthermore, low-rise buildings are commonly constructed of brickwork, which is considerably more sensitive to distortion than a framed structure. Thus, where predicted total settlement of a fill is greater than 20 mm, isolated footings are generally not suitable for brick buildings, and a more rigid foundation is required. In practice, this will probably apply to all cases except where a shallow fill has been properly compacted, or otherwise treated, over its full depth (Section 20.1.4).

Figure 39 *Settlement of a house on a raft (photograph by courtesy of NHBC)*

Design strategy. Where practicable, the economic solution is to design buildings as small independent units, each on its own rigid foundation so designed as to protect the superstructure from effects of differential settlement. To counter the risk of tilt, the design of each unit should be such that the centre of gravity of the superstructure loads occurs in the central zone of the unit plan area. Buildings should be rectangular in plan (with length not greater than twice the width), and with plan area not exceeding about 100 m². Terraced houses, or buildings of greater extent, are not amenable to this treatment and would require some other solution such as piling.

With the exception of shallow fills treated to full depth, the proper foundation for small building units is generally a raft slab of sufficient thickness. The alternative, of beam and slab foundations, can be designed for adequate rigidity (Holland and Lawrence, 1980), but in practice may prove less economic than a plain slab.

Design of semi-rafts. On the other hand, small structures on fill are frequently founded on so-called 'semi-rafts', in which reinforced downstand beams form an edge stiffening to a connecting slab, usually only some 150 mm thick. Such foundations tend to act as tied strip footings and may not always give adequate rigidity to protect the structure as a whole. However, although not strictly rafts, these foundations are clearly stiffer than the conventional pad or strip footings. When properly designed they can be satisfactory in appropriate situations.

One advantage of semi-rafts, on fills of medium to high shrinkage potential, is the effect of the downstanding edge in taking the foundation deeper into the zone affected by seasonal moisture variation.

Suggested minimum requirements for semi-raft foundations for housing include:

* Sufficient internal beams to be provided for the proper stiffening of the beam/slab complex; plan area contained between beams not to exceed 35 m²
* Minimum depth of perimeter and party wall beams to be 450 mm, with some internal beams of same depth as perimeter beams in large dwellings (e.g. greater than 100 m² plan area)
* Beams to be designed to span 3 m simply supported, or 1.5 m as cantilevers, and reinforced in accordance with CP114 or BS8110
* Floor slab to be not less than 150 mm thick, with designed bottom reinforcement and at least nominal reinforcement in the top face.

Reference should also be made to the latest NHBC guide note.

Raft design. The design of true rafts requires an understanding of the principles of soil-structure interaction and relative stiffness. The rigidity or otherwise of a foundation may be assessed using the relative stiffness parameter K_r. For a rectangular raft of thickness, t, and breadth, B:

$$K_r = \frac{4\,E_r\,t^3}{3\,E_s\,B^3} \cdot \frac{(1 - v_s^2)}{(1 - v_r^2)}$$

where E_r, E_s, v_r and v_s are the Young's moduli and Poisson's ratios of the raft and soil respectively. The influence of K_r on differential settlement and bending moment in a raft is illustrated in Figure 40. Although the figure relates specifically to a rectangular raft with a length : breadth ratio of 2, the principle applies to rafts generally.

It can be seen that the transition from a relatively flexible to a relatively rigid system requires a change in relative stiffness of only 100. Since K_r varies as the cube of the raft thickness, the system behaviour is very sensitive to this dimension. For small structures on poor fill ($D < 4$ MN/m² from Table 17) it is desirable that K_r be of the order of 5–10, which is usually achieved with a raft thickness of about 450–500 mm. For better (stiffer) fills the thickness of raft necessary to provide the degree of relative stiffness becomes greater than 1 m. However, the scale of potential movements is such that greater flexibility can be tolerated and rafts in the order of 500 mm thick can provide satisfactory foundations.

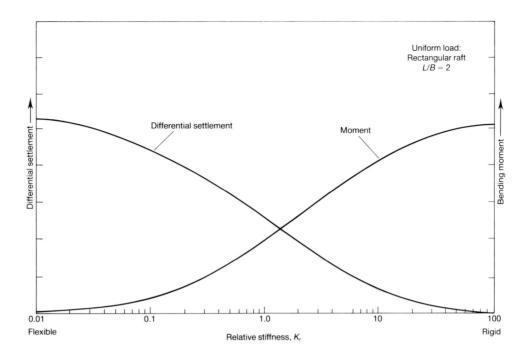

Figure 40 *Relative stiffness of raft, bending moment and differential settlement*

20.1.4 Flexible floating support: pad and strip footings

There may be cases where a raft is a less practicable solution than flexible support by
individual pad and strip footings. Where, for instance, a shallow fill has been subjected to
full-depth treatment a suitably reinforced strip footing may be adequate, without recourse
to a raft, for low-rise buildings. Similar considerations may apply to fill treated with stone
columns which may not necessarily extend the full depth of the fill, although some engineers
prefer to retain the security against differential settlement afforded by a raft.

On medium-depth fills a raft foundation is not economic for single-storey buildings, such as
warehouses or light industrial facilities, with loaded floor slabs of wide extent which cannot
conveniently be broken into small units. One solution is a flexible, fully articulated floor
slab independent of bases for concentrated loads. Provided the ground improvement meets
a specification similar to those outlined in Sections 18.4 and 18.5, the floor may be designed
as uniformly supported by the treated fill. Mesh reinforcement is usually sufficient for the
floor slab, but the design must permit independent settlement. The concentrated loads can
be founded on pad or strip footings or, if too great, on piles.

For structures not sensitive to settlement, the structural loads can be founded on reinforced
continuous footings designed to the principles of relative stiffness discussed in Section
20.1.3.

Structural and machinery foundations sensitive to settlement, and foundations subject to
heavy overturning moments, have to be treated separately. Available expedients include:

- Ground treatment to greater depth than for floor slabs, with suitably tied and reinforced
  foundations
- Piling
- Excavation of fill down to good ground, in the locations of the foundations, and
  replacement with compacted gravel or other granular material, or concrete.

It is important to note that floor loadings on fill, having a load intensity less than that of
the adjacent structural foundations but spread over a wider area, can be the cause of
unacceptable movements of the structure, consequent on the imposition of significant
stresses from the floor into materials at depth under the foundations (Threadgold, 1979).

Floor slabs should not be cast in direct contact with the ground where there is risk of sulphate attack on concrete or disruption through expansive reactions in fill/hardcore. It is also important to note that, for dwelling houses, the NHBC do not accept floors bearing directly on fill thicker than 600 mm; instead they require fully suspended floors. For warehouses, etc. not subject to NHBC restrictions, non-suspended slabs would require protection from potential sulphate attack.

20.1.5 Superstructures

Treatment of fill should always be designed to minimise differential settlement. Notwithstanding this precaution, superstructures of notable extent should be designed as framed structures, preferably of simple, statically determinate construction. Joints should be 'hinged' and of adequate capacity to absorb any displacements imposed by residual differential settlement of the foundations.

In the case history mentioned in Section 20.1.2 (Smyth-Osbourne and Mizon, 1984) differential settlement caused such displacements as to exceed the rotational capacity of the frame hinges, effectively producing a rigid frame with severe overstressing.

For plain unframed buildings such as brick dwelling houses, of materials incapable of resisting tension, ground movements causing hogging are considerably more damaging to the structures than those causing sagging. Expedients to resist the tension caused by hogging (such as ties at wall plate level or reinforcement in the brickwork) are not considered practicable or effective. The essential is a foundation slab sufficiently rigid to resist distortion.

However, despite the assumption of rigidity in the design of foundation rafts, it is good practice for long unframed brick building superstructures to be well articulated by slab-to-eaves openings or frequent vertical expansion joints in unbroken lengths of brickwork (Holland and Lawrence, 1980).

20.1.6 Roads on fill sites

For the design and construction of roads, useful information is given in *Road Research: Construction of roads on compressible soils* (OECD, 1979), insofar as it applies to the UK.

Generally, fill should be so graded that roads may run at approximately the level of the adjacent terrain, and cuttings through fill should be avoided. If roads in cuttings are unavoidable, the cut surfaces should incorporate a granular break/drainage layer with a low-permeability cover. Adequate drainage should be provided in the bottom of the cutting, discharging into an approved sewer or other collector if contamination is present.

Economic considerations generally dictate that some settlement of the finished road surface will have to be accepted. Roads on fill tend to be service access roads, where the riding quality is not critical. Planning should allow for temporary road surfaces, with provision for periodic maintenance and regrading.

If roads without settlement are essential, complete removal of the fill and replacement with sound material may be a competitive expedient in fills up to about 5 m deep, provided adequate backfill material is easily available. Constructing high-quality roads on deeper fills may be very expensive.

For road bases on fill, a low CBR (California bearing ratio) should be assumed, to give a total construction depth of not less than 0.5 m. Care is required to prevent ponding, both on the road surface and in the immediate vicinity. Where roads are to be used also by construction traffic, they should be topped with a temporary carpet approximately 0.5 m thick, with adequate crossfall for drainage. An impermeable geomembrane can prevent water access into the fill under the road sub-base.

Road pavements should be flexible, to facilitate maintenance and repair. Concrete pavements are difficult to remedy in case of settlement damage, and should not be used for roads on fill. Concrete paving blocks and brick paviours have been used on fill in docks and harbour areas.

Embankments, if unavoidable above fill, should be of restricted height. An embankment constructed above existing compacted fill may be subject to:

- Instability of any underlying 'cohesive' fill
- Settlement through continuing creep of the underlying fill under the load of the embankment.

For shallow fill, complete excavation and replacement of the fill below the embankment may prove a feasible solution to both of these problems. For deeper fill other considerations apply, as outlined below.

There is unlikely to be a stability problem with 'granular' fills which have been properly compacted. With 'cohesive fills', however, the extra load of an embankment may provoke instability until excess pore pressures have dissipated. The usual solution is to maintain shear stresses in the fill below the initial undrained strength by;

- Modification of the embankment geometry (reduced slopes or, more economically, loading berms)
- Use of lightweight embankment material such as pfa.

The alternative is to construct the embankment in stages, with continued monitoring of the pore-water pressure in the underlying fill. Under favourable circumstances this may be the most economic solution but may increase the construction time.

A common method for reducing the settlement of the top of an embankment on fill to manageable dimensions is to surcharge the embankment. Between 10% and 20% of the embankment height is an adequate thickness of surcharge, and the method is economic provided suitable surcharge material is available.

20.2 Design against combustion hazard

Proper ground treatment (Section 19.6) is the prerequisite for protection against fire hazard. Design expedients may usefully supplement the ground treatment but are no substitute for it.

Possible design expedients include:

- Providing ventilated foundations for heat sources such as furnaces and incinerators. (PSA have developed a design to protect telephone exchanges from the hazard of gas leaks by placing boilers on the roofs)
- Using lightweight aggregates to increase the fire-resistance of concrete members
- Adapting foundations to serve effectively as 'fire fences' (see Section 19.6.5).
- Adoption of design and construction measures to exclude air from the combustible. Piles should not be used.

In one of the few recorded cases where buildings have been erected on a Category B site (non-spontaneously ignitable material—Section 19.6.4), strip footings were adopted, supported on inert material (pfa) in trenches down to good ground below the fill. Within 24 hours of placing, the saturated (pfa) set hard to a strong load-bearing wall, which formed an effective fire-fence against migrating underground fire (Weatherley, 1979).

20.3 Design of buildings against gas

Before building development is contemplated on a gas-hazardous site careful attention should be given to the reservations expressed in Section 17.8 and any further restrictions imposed by good practice.

Experience in the execution and behaviour of specially designed buildings in gas-hazardous areas is, so far, very limited, but pressure on building space has occasionally led to the exploitation of such sites. The guidance notes given below are based on the experience gained in these special developments, but they should not be taken as guaranteeing approval of such work in the future.

Some of the earlier developments were for dwelling houses, i.e. forms of development which are not recommended in Section 17.8 and may not be acceptable in the future (Waste Management Paper 27, HMIP, 1989). The principles are, however, also applicable to warehouses, supermarkets and similar enclosures, except that free, unforced ventilation may no longer be an acceptable solution.

Protection also has to be provided to surrounding areas such as car parks to ensure that gas is guided to vent in places where it will cause no harm.

20.3.1 Basic concepts

The first essential for the protection of buildings on or adjacent to a gassing site is never to permit a build-up of methane in confined spaces to exceed 20% of the lower explosive limit (i.e. methane concentration in air not in excess of 1%) or safe carbon dioxide levels (0.5%). Design of ventilation systems should give a factor of safety of, say, 2 on these concentrations. There is, however, divergence of opinion as to how that may best be achieved. For example, under the Building Regulations (1985), Approved Document C1/2/3 recommends the elimination of '. . . voids due to settlement of filling . . .'. In contrast, GLC practice made profitable use of a deliberately created underfloor void in the building redevelopment of Surrey Docks, where river silt can emit methane (Pecksen, 1985; Thomson and Aldridge, 1983).

Generally, modern practice favours underfloor voids with gas-tight ground-floor construction. The void acts as a buffer space from which gas is extracted intermittently, either by free ventilation or by forced ventilation actuated by methane sensors, together with a continuous monitoring alarm system. Both alarm and ventilation systems should be vandal-proof and safeguarded against power cuts; they should also be non-sparking and explosion-proof. Systems such as this depend on floor slab integrity in relation to joints, cracks and service entries.

For the design of voids the method described by Pecksen (1985) has proved useful. A void-filling rate is derived from the measured borehole-emission rate, and relates underfloor dynamic drag to void height and ventilation louvre sizes.

For dwelling houses at Surrey Docks free ventilation of the void was adopted. Free ventilation relies on natural air movements outside the building, and ceases during periods of dead calm. Normally these are of short duration in the UK. With free ventilation the sizing of the void must be such that any methane build-up remains below the acceptable limit during the longest predictable calm period.

On larger buildings below-slab ventilation may be provided by a series of gas-drainage pipes, laid in a uniform layer of granular fill or in gravel-filled trenches. These pipes may be passively vented or connected to a mechanical extract system operated continuously or intermittently activated by alarm sensors. In such installations, composite geotextile/geomesh/membrane sandwiches have been laid above the pipes to aid gas drainage.

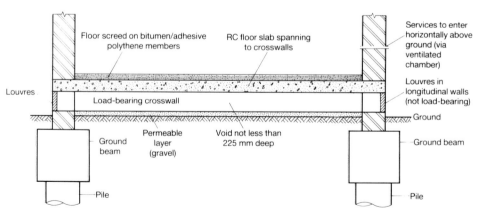

Figure 41 *Cross-section of floor slab showing natural ventilation spaces and louvres (courtesy of London Scientific Services)*

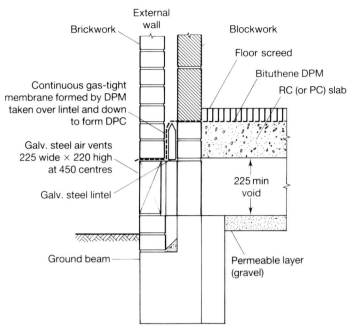

Figure 42 *Vents in external walls of houses (courtesy of London Scientific Services)*

20.3.2 Construction principles

Figures 41 and 42 are based on details from principles established by GLC on underfloor ventilation, but are not presented as construction details. They are intended to illustrate the basic principles discussed in the text.

Current practice requires that the ground-floor slab over the void be as gas-tight as is reasonably practicable. Natural dispersion should then prevent any build-up within the building itself. Buildings in gas-hazardous areas should normally be piled to avoid the risk of cracking the floor slab by uneven settlement. Piles (and vibrated stone columns, if adopted) may form preferred paths leading gas towards the void space. This may not be a disadvantage, since the gas should then emerge at predictable points where building design can more easily deal with it.

For the Surrey Docks scheme the houses had ground beams spanning between pile caps (see Figures 42 and 43). The underfloor void was created by supporting the floor slab on shallow transverse walls of load-bearing brickwork above the transverse beams. The

Figure 43 *Louvres for underfloor ventilation (photograph by courtesy of GLC)*

longitudinal walls below floor slab level, along the front and back of each house, were perforated with louvred grilles throughout their whole length. These louvred walls formed the ventilated sides to the underfloor void, and were not load-bearing. (*Note*: These louvred walls at Surrey Docks were part of the free-ventilation system; forced ventilation would use a slightly modified detail.)

For small buildings the floor slab should be designed to span one-way between the cross walls, with a flat soffit free of downstanding beams which might impede gas evacuation. Where underfloor insulation is required it must also be detailed so as not to interfere with the free evacuation of gas.

Some designers strengthen the slab with downstanding ribs projecting not more than 75 mm, on the assumption that these do not prejudice the proper function of the void. This approach is not proven by long experience and should be used with caution. For wide-span floors in warehouses, etc., deeper downstanding beams may be necessary. These beams should be perforated at frequent intervals by 100-mm diameter holes across the neutral axis (that is, practically at slab soffit level) to permit unhindered ventilation.

Prestressed floor slabs have been used to provide a gas-tight floor (Rys and Johns, 1985). Ideally, a crack-free floor should be provided by a slab cast-in-place under strict control. However, formwork cannot always be recovered from the restricted underfloor void, and is likely to interfere with the designed ventilation. One expedient to overcome this is the use of false shuttering suspended permanently in position by hangers cast into the slab, but this has proved awkward and costly in practice.

The alternative preferred by many contractors is the use of precast slab elements. With this method, careful detailing is necessary to form a seal over the inevitable fissures between individual slab elements.

Figure 43 is a schematic view. Above the precast slab is placed a low-permeability membrane (e.g. bituthene DPM, or 1000-gauge polyethylene bedded in bitumen). On the membrane is formed a floor screed with mesh reinforcement as security against cracking.

In all cases, whatever the chosen method of floor slab construction, it is good practice to supplement the void with a membrane incorporated into the floor slab and walls to impede mass flow of the gas. For this, 1000-gauge polyethylene is generally found satisfactory. Its effectiveness is enhanced if the joints are lapped and taped. Polyethylene is not totally impermeable to methane, but should form an adequate barrier, provided it is not holed. Great care is needed to avoid perforation during installation. Special membranes are now available, such as aluminium foil sandwiched in polyethylene.

The floor (solum) of the void should be as permeable as is reasonably practicable, to encourage gas to enter the void space. It should not incorporate any membrane, blinding or other impedance to upward gas migration. It should normally be formed of the existing ground, possibly topped-off (or replaced as opportune) by a permeable gravel layer.

To maintain the gas-tightness of the floor slab, services should enter the building through the walls above the slab, via an external ventilated box. This may not be possible for drains. Any pipes through the floor slab should be sealed with puddle flanges, and should have sufficient flexibility below the floor to avoid damage to the slab in the case of settlement.

With a ventilated void the ground-floor slab is above the external ground level by about 0.5 m or more, necessitating a ramp or steps to enter a building. This may not be convenient in a warehouse. An expensive alternative would place the void below ground level with a structural trench (or, in the case of forced ventilation, a continuous flue) around the building.

In general, underground structures such as cellars should not be permitted. If unavoidably necessary, they should be designed as gas-tight, with provision for internal ventilation (at least ten air changes an hour), and should be completely surrounded with a granular-filled, venting (gas-drainage) trench.

20.3.4 Further design information

BRE Digests Nos 206 (1977) and 210 (1978) give information on calculation of gas build-up and natural ventilation requirements. BS5925: 1980 also covers the action and design of natural ventilation systems.

20.4 Design against aggressive attack and corrosion

Remedial treatment (Section 19.7) is intended to reduce, as far as practicable, contact between aggressive chemicals and building foundations, services, etc. Thereafter the design option should be exploited to offer enhanced resistance to the potential for aggressive attack remaining in the ground. The design approach operates under three headings:

1. Specification of materials
2. Design detailing and construction methods
3. Protective measures, coatings, etc.

20.4.1 Specification of materials

CIRIA Report 98 (Barry, 1981) gives useful guidance on the durability of many materials in aggressive ground. There still remains, however, much uncertainty as to the quantitative effects of specific concentrations of aggressive chemicals. Given the great variety of building materials (cement-based, resin-based, bituminous and ceramic products, stone, brick, plastics, metals, etc.) likely to be affected, and the many modes of attack to which they may be subject on derelict sites, the engineer should seek the advice of corrosion experts whenever aggressive conditions are identified which are outside normal engineering practice.

A familiar example for engineers in a greenfield environment is the specification of concrete to resist naturally occurring sulphate concentrations in the ground. Numerical values of acceptance levels relative to sulphate attack are quoted in BRE Digest 250 (1981), but are only applicable to aqueous solutions at near neutrality (between pH 6 and pH 9). Dependable values for sulphate aggressivity in the usual acidic conditions of derelict sites do not yet exist. However, Harrison (1987) has recently given data on the durability of concrete in acidic conditions, extended down to pH 2.5, with suggestions for future amendments to BRE Digest 250.

20.4.2 Design and construction

The items most affected under this heading are concrete components cast in place in aggressive ground. Significant corrosion requires the continuing replenishment of the depleting chemical by groundwater migration. Normally this is very slow, but increases notably in situations where evaporation can take place from a free surface (e.g. from a floor or wall slab—Section 12.12). Thus piles and massive foundations, completely immersed on all surfaces, are less affected by sulphate attack than floor slabs cast directly on the ground. The latter are particularly susceptible to attack.

The NHBC do not normally permit the use of non-suspended floor slabs cast directly on more than 600 mm of fill under the slab; floor slabs must be fully suspended. Although this NHBC restriction is mainly on account of settlement problems, it is also a wise precaution against sulphate attack and expansive reactions in the hardcore. For warehouses, etc. not subject to NHBC restrictions, all slabs (floors or walls), cast directly in contact with fill susceptible to severe sulphate risk, may require the additional protection of a membrane properly jointed and sealed.

There is evidence (Arber and Vivian, 1961) that where freshly cast, high-quality concrete (of smooth surface finish) can be exposed to the air for a sufficient time before contact with sulphate, the resulting carbonation renders the concrete surface less permeable to sulphate penetration. Thus quality-controlled, factory-produced, precast pipes are generally more resistant to sulphate attack. This advantage obviously cannot operate for cast-in-place, non-suspended floor slabs, which remain among the most vulnerable of concrete units.

Massive concrete components deteriorate less rapidly than thin sections, especially in walls and floors susceptible to evaporation as above. Dense, carefully worked concrete tends to reduce the migration of sulphate-bearing waters through the component. To that end also, surfaces to be exposed to sulphate waters should have the smoothest practicable finish.

20.4.3 Protective coatings

Whatever other measures are adopted, applied surface coatings are an advisable additional precaution. Available protective material for concrete components include:

- Polyethylene or PVC sheet wrapping with welded joints and protected, if necessary, from abrasion by stainless steel bands
- Butyl rubber-based coatings
- Factory-applied polyurethane for precast units (e.g. prestressed pipes)
- Special resin pipes in place of concrete
- Site-applied epoxy/polyurethane coats
- Extra thickness of concrete surround.

Not all membranes are of proven durability, and care and expertise will be needed in selection. Where laid on the ground, sheeting and membranes should be protected from abrasion by an underlay of smooth cement blinding or sand. Various combinations of the above coatings were used to protect sewage mains and manholes in connection with the Beckton Alps gasworks reclamation, at an extra-over cost of approximately 50% (Netherton and Tollin, 1982).

For other types of coatings, cathodic protection and other safeguards, reference can usefully be made to *CIRIA Report 98* (Barry, 1981).

20.5 Protection of services

20.5.1 General

This section outlines protective measures for water and gas mains, sewers, and power and telephone cables laid in derelict land. The Water Research Centre (1986) deals in detail with water mains, and three of their reports are relevant to the protection of water services in contaminated land (WRc, 1988a & b, 1990).

A variety of materials is currently in use in the UK for service pipes, as outlined in Table 27. Electricity cables are mainly sheathed in PVC; medium-density polyethylene is also used

Table 27 Materials used for service pipes

	Watermains	Sewers	Gas mains
Ductile iron	O	O	O
Grey iron	O	O	
Mild steel	O		O
Concrete		O	
RPM/GRP*	O	O	
PVC	O	O	
Polyethylene			O
Asbestos-cement	O	O	
Clay		O	

*RPM: reinforced plastic matrix.
GRP: glass-reinforced plastic.

(and has better resistance to low temperatures, waterlogged conditions and hydrocarbons in the ground).

Specially excavated, oversize trenches for services, backfilled with inert materials, are today regarded with disfavour by some authorities, who consider them to be possible drainage paths for contaminants. Instead, services should be above that part of the fill which provides the capillary break (Section 19.1.4). This requirement may determine the minimum thickness of the designed cover. Alternatively, it may prove necessary to install services in specially constructed culverts.

Services laid in derelict land can be at risk from:

- Ground movement in fills
- Decaying organic materials
- Corrosion
- Infiltration of toxic materials.

20.5.2 Ground movement in fill

Services should be laid only in Grade I fill. No stormwater soakaways should be permitted and all drainage should discharge beyond the confines of the fill area.

Pipes may have to resist high loads in case of ground movement, and the pipeline should be capable of flexing. Anchored joints may be necessary for pressure pipelines. Several patent flexible anchored joints are available, as well as jointless pipe, and advice should be sought from manufacturers.

With sewer pipes of clay or concrete, rubber or plastic gaskets give some flexibility. Sewers should be laid to falls sufficient to avoid future risk of backfall. In this regard, special care is necessary at the edge of a filled area. Occasionally, for large pipes or where settlement may be severe, special measures may be justified, such as piling or replacement of the fill by imported granular material. Granular materials possess no tensile strength and it may be advantageous to incorporate geotextiles into the backfill under the sewer. It is sometimes beneficial to introduce a relatively thin reinforced concrete strip to bridge local areas where differential settlement may occur, and to construct the sewer or drain above that level, sitting over the strip on a granular base.

Care and attention to detail is required where services enter buildings from general fill areas. Where buildings are on piles, the services can be supported from the ground-beam system, provided that sufficiently flexible connection can be made to the unsupported lines in the fill.

A raft slab may suffer large uniform settlement and precautions should be taken to avoid failure of services at the edges. Provision must be made in the design for suitable flexible or sliding joints to allow such movements in the pipe.

20.5.3 Decaying organic material

Decay of organic material may:

- Increase the corrosion risk, both to metallic and cementitious materials
- Cause heat build-up to the detriment of plastic pipes
- Create voids beneath the services.

The enhanced corrosion risks should be taken into account in the assessment of the protective measures required. Plastic pipes and plastic-sheathed cable should not be used in fills containing putrescible material or other substances likely to attack the plastic. The preferable solution is to remove all organic material and replace it with inert material.

Voids developing beneath services can only be effectively guarded against by building in some spanning capability in the service line (Section 20.5.2).

20.5.4 Corrosion

All derelict land should be regarded as potentially corrosive. Ductile and grey iro be protected by site-applied polyethylene sleeving, with thermoplastic or petrc *171* impregnated tape wrapping for the joints. In the presence of acid wastes th should be wrapped. For steel pipes, cathodic protection is needed in aggressive ground and requires a full, high-grade coating, free of holidays or pinholes.

For precast concrete pipes, manholes, etc. BRE Digest 250 (1980) gives recommendations for protection from sulphate attack in conditions of near neutrality (i.e. between pH 6 and pH 9). For stronger acidic ground conditions (down to pH 2.5), Harrison (1987) gives recommendations based on recent work at BRE (Section 20.4.1).

Possible expedients are protective wrapping, or the use of alternative materials such as RPM (reinforced plastics matrix).

In the presence of hydrocarbons in the ground, both polyethylene and lead sheathing have been given good protection to power cables.

20.5.5 Infiltration of toxic materials

Infiltration into service pipes occurs either by diffusion through plastic pipes or by entry through pipe joints. Organic contaminants (phenols, solvents, etc.) can diffuse through pipe walls or joints without damage to the plastic but with consequent tainting of the drinking water. Plastic pipes, susceptible to diffusion, should not be used in contaminated land for drinking-water piping. See the WRc reports (WRc, 1988a & b, 1990).

Infiltration through joints is unlikely to be a problem with pressure pipes, but with water mains there is always the possibility of a burst, leading to back-syphonage of contaminant through the joints. In all cases, the engineer should check with the manufacturer as to the suitability of the joints.

20.6 Design measures against slags and other expansive materials

Expansion phenomena (Section 11.4) are often associated with sulphate and acid production. Defensive measures may have to consider these latter hazards at the same time as expansivity. They are treated separately in Sections 19.7 and 20.4.

The most certain defence against expansion hazard for all types of building bearing directly on fill is to allow no expansive material under any foundation (Section 18.6). In cases where such materials must be accepted, appropriate design measures are necessary, as below. These are based on established practice, and are applicable as appropriate to the various types of foundations and depths of fill covered in Table 26.

Raft foundations. For small housing units, less than about 100 m² plan area, NHBC have on occasion, accepted semi-rafts bearing on an inert granular 'carpet', 1–2 m thick, overlying the expansive material left in place. The carpet has the effect of distancing the foundation from possible sources of sulphate and of distributing and reducing the intensity of the strains provoked by the expansive materials below. The principle appears sound, and there are no reports of failures.

Piling. This can give assured protection to all types of building on all depths of fill, provided the piles are protected against possible sulphate/acid attack. Pile driving may be difficult through slag deposits. No pilecaps, beams, slabs or other foundation elements should be in direct contact with the underlying ground. Design must include sufficient void space to allow free heave of the ground as well as lateral expansion.

Piers. Individual piers taken down to sound ground are a similar solution to piling, and may be useful in framed buildings of wide extent. They may be competitive for shallow fills.

Suspended floor slabs. Ideally, all the foregoing systems (except rafts) should incorporate fully suspended floors with adequate void space beneath. For house building NHBC require a suspended floor if the depth of fill exceeds 600 mm. If non-suspended concrete floors are used in buildings not subject to NHBC rules (for instance, in warehouses of wide extent) the design should allow provision for movement and consequental repairs and maintenance from time to time, as well as adequate protection from sulphate action as necessary.

Strip footings, whether reinforced or not, are not recommended for use on expansive fills.

21. Monitoring procedures

21.1 Need for monitoring

Individual test results give values of contamination, for example, valid only at the time of observation. A derelict site is, however, a dynamic system in which values can change over an extended period. Fill, for example, continues to settle at a diminishing rate under its own weight, and leachate tends to migrate and change its concentration in consequence of water movement, depletion of biodegradables, etc.

Monitoring is therefore highly desirable and in some cases essential. It should cover such of the following characteristics as may be appropriate to a given site:

- Contaminant migration
- Groundwater levels and quality
- Leachate movement and quality
- Gas emission and migration
- Ground temperatures
- Settlement.

Early monitoring before construction is an essential tool of site investigation which can indicate probable future values of critical parameters and can be vital in decisions on the definitive development of a site.

Continued monitoring after development sounds an advance warning of any change in characteristics likely to constitute a danger, so as to permit the timely institution of countermeasures. In practice, this may mean the timely activation of reserve measures which have already been built into the system (such as pumping from wells to protect the aquifer, when monitoring indicates a deterioration of a protective barrier system—Section 19.8.3).

21.2 Contaminant migration

On sites where known contamination is to be left *in situ*, the engineer should remain alert to the possibility of migration of contaminants whether through groundwater movement, soil suction, uptake by plants or any other cause. Upward migration may put at risk the topsoil or other sensitive targets involved in the development. Lateral migration may pose a threat to adjacent property, while downward migration may lead to contamination of an aquifer.

In all cases where possible targets are considered sufficiently sensitive, early monitoring will be required for assessment purposes and continued monitoring for an indefinite period will be needed to provide advance warning of possible risk.

Monitoring will involve chemical analysis of samples of:

- Topsoil and subsoil
- Vegetation
- Groundwater.

Sampling techniques, the depths of soil sampling and the analyses required should correspond to those given elsewhere in this publication, appropriate to the sensitivity of the target.

Vegetation, both deep- and shallow-rooted, required for determinations of uptake, should be scrupulously separated from the topsoil (with care to avoid washing the leaves and stems). Washed and unwashed samples should then be analysed for contaminants and compared. Much of the metal contamination of a plant may be due to airborne dusts on the leaves. At Southwark (Lyons, 1984) washing was found to reduce by 60% the lead contamination assayed on kitchen-garden leafcrops.

Groundwater samples should be taken from boreholes with sampling tubes left in place. Variation of contaminant concentrations should be plotted against groundwater levels.

Lateral migration will probably be caused by lateral groundwater movements and should be assessed by analysis of water samples from observation wells/boreholes in the vicinity. Where aquifer contamination by downward migration is thought possible, water samples should be taken from boreholes constructed for the purpose. The provisions of Section 15.5.4 apply.

21.3 Groundwater levels and quality

Groundwater levels may affect:

- Settlement of fill
- Emission of landfill gas
- Underground combustion.

Early and continuing monitoring will be required in appropriate cases.

Groundwater levels and how they change should be measured in standpipes left in boreholes, preferably with automatic recording. The records should be correlated with local rainfall and local pumping or other activities likely to affect the water table. All such work should be in accordance with Section 15.8. Redox potential tubes left in boreholes can give information on groundwater quality.

21.4 Leachate movement and quality

Section 12.15 demonstrates the slow progressive changes in the composition of landfill leachate resulting from aerobic/anaerobic action, with the intermediate formation of lower carboxylic acids, followed by reduction into simpler compounds and elements.

Monitoring leachate changes over an extended period can give valuable information on the gradual depletion of the putrescible content of a refuse fill or changes in the contaminant/ soil regime. Where co-disposal of domestic and industrial waste has been practised, the chemistry is somewhat different, and leachate should also be checked for the presence of PCBs, dioxin and other non-biodegradable industrial components. Monitoring should be in accordance with the *WRc Report TR 91* (Naylor *et al.*, 1978).

Continued monitoring also serves to sound a warning of any threat posed by a leachate plume to groundwater resources (Section 19.8.3) to permit the activation of protective measures built into the system.

21.5 Gas emission

Where gas generation is suspected, early monitoring should be initiated as soon as possible. The layout of monitoring points should follow that required for the investigation of landfill gas set out in Appendix 4. It is important to include within the monitoring scheme the immediate area outside the gas-generating fill, particularly when development off the fill exists or is contemplated. Waste Management Paper 27 (HMIP, 1989) suggests that a band 250 m wide around the fill be included in the monitored area.

Whether or not a gassing site is developed, it and the adjacent area will require continued monitoring to demonstrate either the effectiveness of gas-control measures installed or the continued absence of a gas hazard. This is normally effected by continuing to monitor from some or all of the boreholes used for early monitoring.

For the ground monitoring outlined above, detector probes or sampling tubes sealed in boreholes have to be installed (see Appendix 4). Measurements are taken with either portable sensing equipment (Figure 44) or with downhole sensors coupled to an automatic data collection and/or read-out system. A good review of measuring equipment available is given by Crowhurst (1987).

Figure 44 *Monitoring for methane in permanent gas and water observation well (photograph by courtesy of M. A. Smith)*

Buildings constructed on or adjacent to gassing sites will require monitoring for safety purposes in addition to the ground monitoring outlined above. Sampler sensors will be required in ventilation voids and at strategic points in the building where gas can accumulate. These should be linked to an automatic alarm system and the complete installation made vandal-proof.

21.6 Ground temperatures

On sites where the presence of combustible materials is suspected it is important to institute early monitoring. Continued monitoring is required where combustible materials remain after development. Sensing probes inserted in the ground and connected to a central recording point will permit continuous monitoring of ground temperatures. Probes should be inserted on a regular grid and at regular depth intervals to permit the plotting of isotherms in three dimensions. Additional probes should be inserted at points of known elevated temperatures or where combustibility is suspected from appropriate laboratory analyses. Aerial thermal-imaging infra-red sensing is a possible aid to monitoring ground temperatures if carried out on a regular basis.

21.7 Settlement monitoring

For a valid assessment of the geotechnical suitability of a filled site before construction, monitoring of settlement should be started early and carried out for an extended period, ideally not less than 12 months. Continued monitoring after construction is also desirable, especially where a gas hazard may be present. Excessive settlement may lead to cracking of previously gas-tight floors. Unfortunately, the importance of post-construction settlement monitoring is often not fully appreciated.

Recording at regular time intervals the levels of points on the surface is usually sufficient but the levels of points within the body of the fill may also be monitored. Plotting the recorded levels against the logarithm of elapsed time (if possible, since the formation of the fill) should give by extrapolation an indication of the future settlement to be expected and confirmation of performance of the fill.

Levelling is required on a large number of surface-levelling points established on a grid, normally 50 m by 50 m, but with closer spacing where differential settlement may be expected, for example, at zones of rapid change in depth of fill and areas where buildings are located. The datum for the levelling must be outside the area of influence of the settling fill and any similar effects. Cheney (1973) gives guidance on precise levelling.

Surface-levelling points should be secure, robust and vandal-proof. The definitive levelling points for long-term monitoring should be properly designed, of sleeved rods concreted into the bases of shallow holes. As a simpler, interim measure good results can be obtained by using steel rods of 20 mm diameter, 1.2 m long, driven into the ground so that not more than 100 mm protrudes above the surface. Knipe (1979) has found these to give consistent results within 1 mm of those from the more sophisticated definitive levelling points.

Sub-surface movements may conveniently be measured with magnet extensometers, described by Marsland and Quarterman (1974). This comprises a series of ring magnets anchored at intervals to the walls of a 150-mm diameter borehole drilled through the fill with a central access tube for measurement. Typically, the magnets are installed at intervals of 1–5 m.

22. Contractual guidance

22.1 Introduction

This section is concerned with the contracts for reclamation (ground remedial treatment), foundation works and installation of services in the development of a derelict site. Much of the content is also applicable to site-investigation contracts. It discusses the aspects of particular relevance to works on derelict sites which give rise to contractual problems and gives guidance as to how the Engineer can mitigate these problems in putting together and administering the contract.

The preparation of contracts for works on a derelict site, especially one suffering from contamination, is a more complex procedure than for a greenfield site. It requires a proper understanding and fair apportionment of the risks and a clear declaration on the allocation of responsibilities of the parties, and financial liabilities in case of inadequate performance.

Once the ground-treatment and foundation works are completed, construction of works above ground is subject to considerably less uncertainty and risk and is amenable to a normal building type of contract. It is thus feasible, and in most cases desirable, to have two separate contracts: one for the works to ground level and another for works above ground.

22.2 Contracts for work to ground level

22.2.1 Type of contract

Site investigation, carried out as rigorously as is economically feasible, is still limited in what it can reveal. On a derelict site, there is always the possibility of discovering unsuspected contamination or other underground hazard during the execution of ground remedial treatment or foundation works. This may alter the scope of the contract, or at least provoke additional works. Thus the scope of the contract may not be fully definable at the tender stage.

The Engineer should ensure that the Employer (developer) is aware of the very real risk of encountering additional hazards—a risk which cannot be passed on to the Contractor. The overall budget should allow for such contingency. Any attempt to require the Contractor to absorb the effects of such unforeseeable hazards would be unfair and result in an unrealistic and uneconomic tender price. The contract should therefore be sufficiently flexible to allow for such modifications as may be enforced by hazards brought to light during its performance, and to provide means whereby the Contractor may be paid fairly for carrying out the additional work.

The types of contract, generally in use for building and civil engineering works, include:

- Lump-sum
- Cost-reimbursable
- Remeasurement.

The lump-sum contract is convenient to administer, but is only feasible when the full scope of the work can be defined at the time of tendering. It is generally inappropriate for civil engineering works below ground level and is even more so for remedial works on a derelict site. Given the impossibility of definition of such works, the inflexibility of a lump-sum contract would leave the tenderer no alternative but the protection of quoting an excessively high tender price. This type of contract is, therefore, not recommended for reclamation, foundation works or for site investigation works on a derelict site. It would, however, normally be suitable for the building and construction works above ground level, where the scope can be sufficiently accurately defined.

Cost-reimbursable contracts allow a certain flexibility, but they are not liked because costs are difficult to control. Forecasting of the final contract cost and realistic budgeting are rarely possible. Administration of this form of contract is difficult and open to abuse, and this type of contract is not recommended.

Contracts based on remeasurement are the best suited to remedial and foundation works on derelict sites as well as site investigations. Depending on the form of contract chosen, they can provide sufficient flexibility for the protection of both Employer and Contractor and they are the most likely to provide a realistic tender price. Bills of Quantities prepared by the Engineer, from information gained during the site investigation, allow a fair tender comparison and form a basis for budgeting.

22.2.2 Form of contract

Of the forms of contract in common use, the Institution of Civil Engineers Conditions of Contract, 5th edition (1973) is the best suited to works below ground on derelict land. It is a remeasurement contract and its Clause 12 (adverse physical conditions and artificial obstructions) provides the necessary flexibility in case of discovery of unforeseen hazards. It is also suitable for buildings and superstructure works.

The Department of the Environment Form of Contract GC Works 1 does not accept the possibility of a Clause 12 situation and, although a remeasurement contract, is not entirely suited to the works below ground necessary on a derelict site.

Other forms, such as RIBA and JCT, are designed for building work and are generally considered not sufficiently flexible for normal works below ground. They are even less appropriate for the special conditions below ground on derelict sites.

22.2.3 Content of contract

The Engineer has to exercise considerable care in drawing up the contract documents to ensure:

- Realistic tender rates
- Safety of site workers
- Protection of the environment and neighbours
- Continuing monitoring of site conditions during all investigation and construction works
- Proper functioning of the completed works
- Proper appreciation of the risks inherent in the site, and the need for adequate protective measures
- Clear understanding of the positions of the parties in the event of unforeseen hazards coming to light, and actions to be taken.

The documents should be as informative as possible so that tenders may be submitted in full recognition of the known facts. The Engineer should review the documentation critically to ensure that it covers all possible consequences likely to ensue from unexpected hazards.

Realistic prices, resulting from attention to the foregoing details, take account of known facts, and provide a clear basis for consideration of claims and payment should unexpected hazards come to light.

When considering the Conditions of Contract, the Engineer should pay particular attention to the following.

Sureties (Clause 10, ICE 5th edition). The Performance Bond should be set somewhat higher than in normal civil engineering work, in view of the cost to the developer in the event of the Contractor's failure to perform in the face of unexpected hazard. Abandonment by the Contractor may well leave the site in a more dangerous state if a previously tranquil contaminant, for example, has been rendered reactive through disturbance.

Suspension of work (Clause 40, ICE 5th edition). There may be motives for suspension by the Engineer, beyond those operative in normal works, such as intervention by the local authority to prevent pumping of contaminated water from the site. The Engineer should consider whether the normal 90 days' suspension is sufficient.

Programming of work. A clause should be included which gives the Engineer power to dictate the order and detailed scope of work to suit findings during the progress of the works.

Record drawings. Engineering contracts require the Contractor to produce record drawings of the completed works. This is a particularly important obligation on a derelict site because it has serious implications for the safety of future maintenance workers, occupants and others. Record drawings should show the information called for in Section 23.1 in sufficient detail to permit future excavation and other works without incurring risks from the hazards already known from discovery during the development.

22.3 Specification

22.3.1 Requirements

The Engineer can do much to overcome the particular problems of reclamation work on a derelict site by producing a clear and unambiguous specification. The contents should be carefully reviewed to ensure that it covers all reasonably foreseeable situations which may occur during the works. It should include detailed courses of action to be followed on encountering unforeseen hazards and should specify the provision of certain equipment in anticipation, such as tanks for storage of highly contaminated liquids.

Given the limitations of site-investigation procedures, some hazards may only be suspected without hard evidence from the factual data provided to the tenderer. Inferred conditions should be anticipated in the specification.

The specification should also indicate clearly the constraints which will be put on the Contractor by outside parties, such as water authorities, and should specify the criteria which they will apply to the operations. Although tenders will be sought only from contractors experienced in remedial works, the specification should be adequate as a guide to those who may be unacquainted with the specific demands and hazards of the derelict site in question.

22.3.2 Type of specification

The preferred type of specification is one where method is specified rather than performance, because this places the onus on the Engineer and the multi-disciplinary team to carry out all necessary design and research. It also produces tenders which are comparable at the assessment stage.

The various specialist processes which may be involved require equally specialist input to the specification. Given that the Engineer's team has been appropriately constituted for the investigation and appraisal stages, there should not be a problem in specifying by method.

It is relatively common, however, for some types of work to be specified by performance. Ground treatment, piling, and protective coatings come under this category. This approach is open to unequal tendering and leaves the Engineer with the task of assessing tender proposals which are not comparable. To do this properly, the Engineer must have sufficient knowledge of the design and specialist process or material. If that is the case, the Engineer should have been able to specify by method from the outset.

Performance specifications often call for or attract bids accompanied by a guarantee. Whilst this is thought by some practitioners to overcome the potential problems, it should be remembered that liability under such guarantees is notoriously difficult to prove and is unlikely to cover full consequential damage.

22.3.3 Safety

The prime responsibility for the safety of the works, site workers and third parties rests with the Contractor, and the Engineer should be careful not to assume some of that responsibility through overspecifying. Nevertheless, the Engineer must at all times be satisfied that the Contractor is fulfilling contractual obligations, and that the Contract is being performed in a safe manner and without the creation of nuisances which might involve the Employer in litigation. Close collaboration between the Engineer and the Contractor can do much to promote safe working on-site.

Section 23 sets out the minimum standards required for safety measures (facilities, clothing, equipment) and procedures for site working. The contract must ensure that these standards are maintained throughout the performance of the contract.

Basic principles and procedures for safe working should be written into the contract, on the basis of information gained from the site investigation. The specification must clearly assert the high standard required and should state the safety measures, procedures and controls which will be required of the Contractor. These will be considerably more onerous than in normal engineering construction. A clear specification makes for fairer tendering, ensures that safety-conscious tenderers are not penalised, and alerts all tenderers to the chemical and other hazards on the site.

If the site investigation has revealed severe site hazards, it may be convenient to specify the use of a specialist contractor to reduce these hazards before the start of the main site works.

The contract should require the Contractor to nominate the Safety Officer(s) whose duties shall include the policing of the works and ensuring that safety standards are not violated. The expertise required under this heading is set out in Section 23.5.2. All advance testing for the safety of site operations should be done by the Contractor, but the contract should include the right of the Engineer to check the results. Instructions can then be issued to improve safety if necessary. To that end, the Engineer will need to have access at all times to the requisite expertise on site. If there is insufficient improvement, the Engineer may have to use the power under the contract to suspend the work.

22.4 The tender

22.4.1 Invitation to tender

Only qualified contractors, with proven experience in working on derelict sites, should be invited to tender. A select few companies already have recognised status in such work, as, for example, the principal companies in the UK offering specialist ground-treatment processes. If it is desired to give opportunities to less established companies, pre-qualification is advisable, following the normal procedure for civil engineering works.

With the invitation to tender, the Engineer should issue documentation sufficient to cover the following:

Condition of the site—description of site, means of access, etc. as well as information as full as possible on the factual condition of the site, as ascertained from the site investigation, with a clear indication of any uncertainties in the data. This information should be warranted as being factually correct under the contract. It is highly advisable to make the Contractor party to all available factual information from the outset. When warranted under the Contract, it provides a clear basis for consideration of claims.

Bill of Quantities—sufficiently comprehensive to permit realistic comparison of tenders, which should include provisional sums, provisional items or rate-only items to cover foreseeable contingencies (for example, rates to cover work that may prove more difficult than foreseen such as deeper excavation requiring timbering, dewatering, etc.; transport to licensed dump; protective measures against gas, fire, leaking contaminants, etc.).

Constraints—known or likely to be imposed on the Contractor (for example, restriction by local authority on pumping contaminant into foul sewer).

22.4.2 Information to be submitted with tenders

Tenderers should be required to submit documented essential information. It should be made clear in the invitation that non-submission may disqualify. This information should include as a minimum:

- Outline working statement to illustrate tenderer's proposals (to be followed by detailed statement after award of contract). This should be in sufficient detail to show that the tenderer has based the tender on a full knowledge of the available facts
- List and description of plant the tenderer proposes to use
- Proposals for safety of the works, site personnel and third parties. These should include particulars of the intended Safety Officer(s); description and details of manning of site medical centre; statement of site safety policy to show how the tenderer would meet the obligations regarding safety facilities, clothing, equipment and working procedures
- Proposals for containment of contamination, dust suppression, etc.
- Proposals for the assessment and control of contaminated material for off-site disposal
- Layout plan of site offices, changing facilities, etc.
- Provisions for preventing transfer of contamination off-site by transport or workers
- Evidence that known risks have been declared to the insurer
- List of key site personnel and relevant experience
- Evidence of the tenderer's experience in derelict land reclamation works, and technical and financial strength, unless pre-qualified
- Method statements for specialist works, with details of resources to be provided.

22.5 Supervision of construction

The Engineer must appoint site supervisory staff with care and with due regard to the operations to be carried out by the Contractor. Many specialised activities may be involved, and it is of little use to call for the Contractor to be experienced in these if the Engineer's staff on site are not in a position to control and supervise the work from a suitably experienced background. It is to the site staff that the Contractor will turn if unexpected conditions are found. They must be able to react both quickly and positively.

The Engineer will require staff with appropriate basic knowledge and experience, including disciplines other than civil engineering. The inclusion of a chemist or environmental scientist and a geotechnical engineer will be basic to most schemes. On filled sites to be treated, the geotechnical engineer should be knowledgeable and experienced in the process to be used. Other professionals who may be required include hydrogeologists, botanists and corrosion specialists, depending on the specific aspects of the work, but these may be used on a visiting basis rather than as permanent members of the site staff.

23. Safety in site working

23.1 Introduction

Safety of site workers (and of third parties affected by the development works) must be considered at all stages. Safe execution of the works is an essential feature of the design, and the safety provisions required of the contractor should be clearly defined in the contract. Safe working will require suitable protective measures (facilities, clothing and equipment) and the imposition of correct working procedures.

The safety of site personnel is the responsibility of the contractor, and the engineer should do nothing to assume that responsibility. Nevertheless, the engineer must continue to be satisfied the contractor is fulfilling all safety obligations. Close collaboration between the engineer, the contractor's site manager and the local HSE inspector will help to ensure that the available expertise is used for the benefit of those most at risk at the working face. Note that HSE is currently drafting a guidance note on the protection of personnel and the public during development of contaminated land.

In addition to the precautions prescribed below, protection of third parties including trespassers and vandals will require site fencing and other routine works outlined in Section 7.4. Great care must also be taken not to allow dangerous contamination to escape, either off-site or onto 'clean' areas of the site itself.

For the safety of future maintenance workers, careful records should be kept throughout the project, covering:

- Location (extent and depth) of all known or suspected hazards
- Specific locations of samples and details of analyses
- Locations of remedial works and contamination replaced by 'clean' material
- Locations and details of covers and hazards left *in situ*
- Any other information likely to affect future safety.

In addition, records should be kept, at each stage, of all discovered factors likely to affect the safety of working in the successive stage. Thus site reconnaissance should note likely risks to the site investigation team, etc.

23.2 Stages of development and restricted areas

The effective severity of hazards to site workers will depend, in part, on the intensity of the hazard unleashed by the works and, in part, on the awareness of the person at risk. These two factors will in turn depend on the stages of the development reached and on the part of the site under consideration. Certain parts of the site (called Restricted Areas below) may be sufficiently hazardous to require special precautions, whereas other areas may impose far less restraint on the safe movement of personnel.

23.2.1 Stages of development

As regards the intensity of hazards likely to be exposed at any time, and the workers' understanding of the related risks, the development may conveniently be divided into stages, as follows.

Stage 1. Site reconnaissance—where a few skilled persons are subject to hazards which may be known only vaguely from the desk study. The inspection is usually confined to the surface, and the intensity of the hazard exposed at that stage will generally be low. By profession, such site personnel should be very aware of the implications of the site hazards and sophisticated safety devices are not usually called for. Provided work is organised to reduce to the practicable minimum the time of exposure to site hazards, the 'basic' measures of Section 23.4 should be sufficient.

Stage 2. Site investigation—where a larger number of similarly aware specialists and technicians will be exposed to still relatively unknown hazards of somewhat greater intensity, including ones below the surface. Given proper pre-planning of the work and the imposition of safe working procedures, the basic measures of Section 23.4 should be sufficient, with the addition, as necessary, of temporary facilities for washing, eating, clothes changing and first-aid, and sufficient self-contained breathing equipment, harnesses and ladders for the rescue of workers from pits, etc.

Stage 3. Site reclamation—where a large number of unskilled persons will be exposed to hazards of increasing intensity, especially in excavation works, as more of the contamination or other hazards are brought to light. These workers may have little awareness of the dangers and, through indifference or bravado, may be seriously at risk. Very stringent procedures must be followed, and the full range of 'special' measures of Section 23.4 may on occasion be called for.

Stage 4. Building construction—where a large number of similarly unskilled persons will be on the site. However, most of the hazards should have been removed, reduced or covered over in Stage 3. The facilities of Stage 3 (washing, clothes-changing, etc.) may still be required on-site, but otherwise the basic measures of Section 23.4 should be sufficient. If, on the other hand, deep excavations are still required where hazards have been merely covered over, the special measures may also be required.

Stage 5. Post-construction maintenance—where the site should be considered safe for the permanent occupants. Maintenance work will, however, require excavations from time to time. Ideally, the remedial works of Stage 3 should have been adequate to permit future maintenance without danger, so that the basic measures of 23.4 should be adequate (but see Section 23.1 on records).

23.2.2 Restricted areas

Generally, a derelict site will not be uniformly hazardous throughout its entire area. In most areas there will usually be no constraint to the free circulation of personnel. In such areas it will usually be sufficient for personnel to adopt only the 'basic' clothing and safety equipment of Section 23.4 at any stage in the development.

In other areas local hazards may impose serious constraints on personnel movement. Such 'Restricted Areas' should be indicated in the field by signposting, plastic tapes or other visual devices, and entry should be strictly controlled by a permit system (Section 23.5.3). Generally, excavations, pits and confined spaces subject to gas hazards should be classed as Restricted Areas, as well as surface or underground areas containing unstable structures, dangerous liquids, old tanks, sumps or piping, or posing a gas or combustion hazard or underground void, or other areas considered to represent an unacceptable risk without effective countermeasures.

In Restricted Areas the hazards may be so intense as to require sometimes, but by no means invariably, some of the 'special' measures (clothing and equipment) of Section 23.4.

23.3 Hazards and precautions

Sections 11 and 12 describe in some detail the hazards likely to be encountered on a derelict site. The following notes are a resumé of specific aspects posing threats to the safety of site working and of relevant precautions.

Hazards likely to affect site workers include:

Contamination:	Chemical and biological (including radioactivity, asbestos, carcinogens, toxic gas, etc.)
Fire/gas/explosion:	Flammable organic residues, sulphur, carbonaceous minerals and other combustibles; biodegradable domestic waste, decaying river mud and silt
Physical:	Excavations, abandoned pits, lagoons, unstable ground, unsuspected voids
Structural:	Unstable buildings, overhead structures, underground sumps, tanks.

23.3.1 Chemical hazards

Solids. Workers can be mainly at risk from contact or from inhalation of dried-out dusts. Water spraying can help alleviate the latter hazard. During site investigation works where heavy metal residues are suspected, suitable respiratory equipment should be worn whenever pits or trenches are dug or sampled under dusty conditions.

Liquids. All liquids should be presumed toxic, corrosive, flammable or liable to release dangerous vapours, until proved otherwise. Liquids tend to migrate and are more difficult to handle than solids. If allowed to escape from drums, pipes, etc., they can damage a disproportionately large volume of solid ground. For example, PCBs (from transformer oils, etc.) are known carcinogens; they are not biodegradable, and spillages can persist in the ground. Breaching of containment may create hazards through mixing of reactive contaminants. Acids, for example, may react with buried metals, releasing hydrogen, hydrogen sulphide, or toxic oxides of sulphur or nitrogen, depending on the nature and concentration of the acid.

Liquids can be corrosive to the skin and damaging to the eyes. Safety goggles and impermeable clothing may be needed for site workers. Temporary tanks must be provided for the emergency storage of dangerous liquids. (See also Sections 13.1.3 and 26.6.2.)

Gases. These can accumulate in pits, trenches and other insufficiently ventilated confined spaces, and can render excavation works dangerous. Where a gas hazard exists, all excavations, trenches or pits deeper than 1.2 m must be purged and re-tested before work is started or resumed. Excavations of wide extent have been known to fill with gas overnight on a falling barometer (Figure 45).

Figure 45 *Tar tank below ground (photograph by courtesy of Environmental Safety Centre, Harwell Laboratory)*

Gases are usually found diluted with air, and the positive detection of gas may not imply immediate danger. Measured concentrations should be compared with the Occupational Exposure Limits (OEL) published by the HSE (1989).

Sense of smell is important in warning against gas, but is not infallible. Many dangerous gases (such as pure methane, carbon dioxide, hydrogen, carbon monoxide) are practically odourless. To many people, hydrogen cyanide has so faint a smell as to pass unnoticed on a busy site. The very poisonous hydrogen sulphide has a smell easily detected on first contact, but is particularly dangerous because continued exposure dulls the sense of smell. This gas can accumulate in open spaces overnight in still air, and has caused fatal accidents on waste tips.

Gas hazards include:

- *Toxicity*:   for example: hydrogen sulphide, carbon monoxide and hydrogen cyanide
- *Anoxia*:   oxygen may be consumed by smouldering carbonaceous matter, rusting metals or bacterial activity
- *Narcosis*:   from organic vapours, benzene, fuel gases, etc.
- *Asphyxia*:   for example, carbon dioxide, nitrogen, argon
- *Fire/explosion*:   for example: hydrogen, methane, gases from LPG containers and aromatics from petroleum spirit. Flammable gases may necessitate the use of flameproof equipment and great care in drilling operations.

23.3.2 Biological hazards

Soils harbour human pathogens, including bacteria, fungi and viruses associated with sewage sludge or waste/stagnant water. Anthrax spores may be present where infected animals have been buried. Most pathogens are destroyed in a few weeks by sunlight and oxygen in the immediate surface, but excavation can expose personnel to surviving pathogens from the interior of the soil mass: anthrax spores can persist for decades.

Protective clothing should at least include the 'basic' items of Section 23.4 and should be regularly washed and disinfected. Gloves should be rubber and should be regularly inspected and replaced if damaged. Strict personal hygiene is essential, with prohibition on smoking in suspect areas. Site personnel should be immunised to the extent practicable, and instructed to report any unusual symptoms. Where anthrax or other pathogenic spores are suspect, advice should be sought from the Ministry of Agriculture, Fisheries and Food.

23.3.3 Radioactivity

Where radioactivity is suspected, advice must be sought in the first instance from the National Radiological Protection Board, Rowstock, Oxon. Radiation monitoring is needed to locate radioactive contaminants, and should be conducted by trained specialists, properly equipped and protected. A site cannot be declared safe after a simple surface scan with a hand-held monitor. Where radiation monitoring has located radioactive contaminants, advice must be obtained from NRPB before any attempt is made at site remedial works.

Suitably trained investigation personnel may work on a site contaminated by radioactivity, so long as exposure to radiation is monitored and proper health physics support is provided, together with impermeable, washable protective clothing and suitable equipment. In addition to limiting exposure to radiation, great care must be taken to avoid the entry of radioactive substances into the body, where they may cause severe, acute and chronic damage. Strict observance of the Ionizing Radiation (Unsealed Radioactive Substances) Regulations 1968 is essential. These define precautions for protection from risks of inhalation, ingestion or skin contact.

23.3.4 Asbestos

Airborne asbestos fibre is an increasing and unpredictable threat on many types of derelict site, and its suspected presence will call for specialist advice. The hazard arises from inhalation of airborne fibres and, at present, it is not possible to set a measurable limit of exposure which does not carry some residual risk.

Disposal of asbestos is controlled by statutory regulations (Statutory Instrument SI 1983, No. 1649). Before site work commences advice should be sought from the local HSE.

Care should be taken not to allow suspected surfaces to dry out, and to control carefully all excavations liable to bring buried asbestos to the surface. Site work should be organised to keep to the practicable minimum the exposure of personnel. Monitoring of atmospheric fibre concentrations may be needed to check exposure levels suffered by workers, and to give warning of significant dust generation. Where intensive contact with asbestos residues is necessary, full body-cover protective clothing and positive-pressure respirators should be used.

23.3.5 Combustibility hazards

Until a suspect site can be declared safe it should be treated as if potentially combustible. Personnel should be warned not to light fires or take other action calculated to heat flammable materials on the surface or underground. On actively gassing sites, smoking and the use of naked lights should be forbidden. Given the extreme tenacity of underground fires, carelessness may place the developer under a long-term liability, and may even enforce abandonment of the project. The development plan should allow for the immediate, unhindered intervention by the fire brigade before any fire can take hold underground.

The principal hazards are poisonous/asphyxiating gas emissions, and formation of cavities into which personnel may fall on disturbance of the surface. Subsidence may weaken structures and fracture mains.

Protection of personnel will require frequent monitoring of the atmosphere and underground temperatures. Self-contained breathing and resuscitation apparatus should be immediately available. Personnel should never work alone or on a heated area. Where combustion cavities are suspected, personnel should use safety harnesses and walk on planks.

Section 19.6.6 gives advice on dealing with underground burning. Water should never be used to cool hot material underground, without expert advice, due to the risk of generating flammable water gas.

23.3.6 Excavation hazards

Workers are probably at greatest risk during the excavation works required for site reclamation (Stage 3, Section 23.2.1). Large-scale disturbance of the site can lead to widespread hazardous conditions, in which personnel find themselves confined in an aggressive environment, in contact with contaminated ground and possible accumulations of gas, in addition to the risks of unstable sides to excavations. It is this excavation phase which makes the greatest demand on protective material resources and disciplined working procedures. (See Section 23.5.4 on precautions.)

23.3.7 Dangerous structures

Derelict industrial sites frequently contain abandoned gantries, pipework and racks, chimneys and similar overhead structures in dangerous condition. Buried tanks, sumps and pipe runs are underground hazards which may contain harmful liquids.

Dangerous structures and the like should be demolished or otherwise safeguarded, if possible before the site investigation, provided this does not prejudice the accuracy of the investigation. Demolition should be by experienced personnel, preferably by a specialist contractor, and in accordance with BS6187 (BSI, 1982).

23.3.8 Miscellaneous hazards

Tarry materials. These frequently contain carcinogens. Skin contact with these materials, in liquid or in powdered form, should be avoided by the use of appropriate protective clothing. Clothing contaminated by tar should be discarded and not re-used until properly laundered, to avoid risk of continual rubbing of the contaminant into the skin.

Explosives. These may well be unstable. The police should be notified, and the site evacuated until the explosives have been dealt with by experts (Figure 46).

Biodegradation of landfill. Landfill may exhibit any or all of the foregoing hazards to site workers. In addition, the biodegradation of putrescible waste can lead to the formation of dangerous, unsuspected internal voids in an apparently stable material.

Lagoons. Areas of old lagoons, having an apparently firm solid surface, should be treated with caution. The surface may be merely a crust of dried-out material over a still-liquid mass.

23.4 Protective measures

In addition to safe working procedures (discussed in Section 23.5), site safety also requires protective measures, involving facilities, protective clothing, monitoring equipment and safety equipment, as well as pre-planned contingencies for dealing with unforseen hazards.

23.4.1 Facilities

For the health and safety of the large workforce involved in the site-reclamation and construction stages, special facilities should be provided, including:

• Clothes-changing rooms
• Washroom
• Eating room
• First-aid station.

These facilities should be located upwind of site contamination and should have entrances and windows facing away from the site if practicable. Similar facilities may be required for the site-investigation stage, but may be of a temporary, less elaborate nature.

Clothes changing. On badly contaminated sites this facility should have a 'clean' and a 'dirty' section, arranged so as to prevent site working clothes and footwear being taken off-site. Lockers should be provided in each section for the separate storage of workers' off-site clothing and on-site protective clothing. Arrangements should be in hand for the regular laundering of protective clothing.

Washroom. Personal hygiene is important on a contaminated site, and workers should pass through the washroom before entering the eating rooms. In addition to the normal site washing facilities, drench showers should be installed for emergency decontamination.

Eating room. Eating on site should be forbidden except in the eating room, for the safety of all personnel. Muddy boots should not be brought into the eating room, due to the risk of mud drying out into dangerous dust.

23.4.2 Protective clothing

Basic requirement. For day-to-day working in the non-restricted parts of a site, fairly simple protective clothing, of a type comfortable to wear for long periods, should generally be sufficient. For the initial site reconnaissance safety boots alone may be acceptable. For other stages the basic minimum should include:

• Overalls
• Safety boots
• Safety helmets
• Gloves and barrier creams.

Overalls may typically be of terylene/cotton mix, close fitting at ankle and wrist. They should be laundered at regular intervals or whenever contaminated with asbestos, tars, oils or solvents. Workers should be warned of the danger of flammable liquid soaking into clothing; overalls which have been splashed should be changed immediately. Boots may conveniently be of rubber with steel toe-caps. Rubber is not durable in the presence of certain organic liquids, and boots should be inspected regularly and replaced if damaged.

Figure 46 *Gun barrels found in Woolwich Royal Arsenal development. Old munitions sites pose hazards even though these reject castings did not (photograph by courtesy of GLC)*

PVC-impregnated gloves are resistant to many, but not all, common contaminants. They are, however, awkward to wear and lead to loss of manual dexterity, and are likely to be discarded by site workers. A desirable alternative is to make the fullest use of mechanical plant to reduce the need for handling contaminated materials. Barrier creams should always be applied at the start of each working shift.

Special requirements. For the more onerous conditions in Restricted Areas additional protection may sometimes be needed for short periods. Depending on the severity of the hazards, chemical-resistant, cleanable boiler-suit type of garments in PVC may be necessary. For work in great hazard (e.g. radioactive areas, or in atmospheres containing asbestos dust) impermeable protective garments covering the whole body may be needed together with self-breathing equipment. (Note that work becomes progressively more exhausting, the greater the protection provided.)

23.4.3 Hazard detection and monitoring equipment

For the protection of working parties gas-detection equipment will be required, especially in excavations. This is purely a safety operation, as distinct from the accurate gas sampling (Appendix 4) required for site-appraisal purposes. As a basic minimum for site reconnaissance and site investigation, detection of oxygen deficiency and flammable gas may be sufficient protection, provided the results are produced rapidly.

For prolonged exposure of workers during site reclamation in hazardous locations, especially in excavations in Restricted Areas, automatic gas detector/monitors with multiple gas sensors, fully weather-proofed and with audible alarms, should be installed near the working face.

23.4.4 Safety and rescue equipment

Basic requirements. As a simple protection for workers, safety goggles and plain dust masks (or respirator-type face masks with spare filters) may be advisable on occasion. They are not without some inconvenience in wear, and are not always necessary. Face masks are not effective in oxygen deficiency, and should not be relied on as protection against asbestos dust.

Special requirements. Whenever work is planned for a Restricted Area, rescue equipment must be available in readiness at the work face. This should include:

- Breathing apparatus
- Safety harness
- Ladders
- Fire extinguishers
- First-aid box
- Audible alarm (hand-operated).

Even in cases where only simple basic personnel protection is provided for everyday use, sophisticated rescue equipment is justifiable where unforeseen emergencies can arise.

Breathing apparatus may be of positive-pressure self-contained type with back-pack air cylinder, or may be of remote type fed by pipe from a separate source. A lightweight breathing apparatus should be available for the use of the victim to be rescued. For the selection of respiratory equipment, reference should be made to BS4275: 1974, *Recommendations for the selection, use and maintenance of respiratory protective equipment*, and also to the Health and Safety Executive's approval Lists Nos F2486, F2501 and F2502.

A safety harness is required to protect anyone entering an excavation on a rescue mission, or when dangerous conditions are suspected.

The correct extinguishing equipment should always be to hand. Water used on flammable liquid fires can be dangerous. Foam, dry-powder or carbon-dioxide extinguishers are all suitable for small quantities of flammable liquids.

In Restricted Areas a first-aid box should be at the work face, irrespective of the existence of a fixed first-aid station on the site. The size and content of the box should conform to the recommendations of the Health and Safety at Work booklet HS(R)11, *First-aid at work*. On severely hazardous sites, a stretcher and blankets should be available at the work face.

Hand-operated audible alarms should be available to give warning immediately to other site personnel in case of emergency. Alarms may be portable and/or installed permanently at easily visible strategic points. It is often convenient to mount them on an excavator with a flashing light signal. Hand-operated alarms should be differentiated by tone or signal pattern (continuous, intermittent, modulating, etc.) from the automatic alarms of the continuous gas monitors (Section 23.4.3).

23.4.5 Pre-planning and contingencies for dealing with discovered hazards

Planning for site investigation. During site investigation considerable information has to be obtained regarding contamination and other hazards, but the value of the investigation contract is unlikely to justify sophisticated safety measures, equipment and procedures. Site-investigation work, therefore, should be so pre-planned as to reduce to the practicable minimum the time of exposure of personnel to the site hazards. To that end, essential documentation should be prepared in advance, to facilitate the rapid collection, identification and classification of samples.

Contingency plans for reclamation works. However thorough the site investigation, the possibility always remains that the reclamation works may expose unexpected hazards requiring immediate countermeasures, for which contingency plans should have been prepared in advance. If the hazard is severe or of wide extent, specialist advice may be required.

Countermeasures may require removal of additional quantities of contamination, provision for urgent storage of spilt liquids, emergency measures to protect the environment or water resources, etc. Unexpected contamination, once unearthed, may be in a category of Special Waste and its re-burial on site may be prohibited. In other cases, the contaminant may be moved and re-buried elsewhere on site. Full protection will then be required for site workers. Thus, if asbestos comes to light in an excavation, surface water spraying will be necessary, with atmospheric dust monitoring and, possibly, full respiratory protection.

The contract should recognise the need for contingency plans in such cases, and sufficient funds should be in reserve in the budget. The alternative may be a drastic revision or curtailment at a late stage in the development.

23.5 Safe working procedures

23.5.1 Training of site personnel

Reclamation of a derelict site will expose a large number of site staff and workers to hazards with which they may not have previously been familiar. Safe site operations will require proper working procedures and personnel capable of executing them. Prior training is of vital importance in acquainting all concerned with the types of hazard to be expected and procedures for combating them. Before any work starts in a hazardous area the Engineer and the Contractor should be sure that their personnel are sufficiently trained to confront the expected hazards.

Training should cover, *inter alia*:

- Recognition of various site hazards and locations of Restricted Areas
- Proposed site-working procedures and permit system for Restricted Areas
- Location of rescue services on-site, first-aid, etc.
- The need for everyone to be alert and to report unusual conditions
- Use of standard safety equipment, monitoring equipment and rescue appliances
- Emergency drills and alarm systems; methods of rescue
- The importance of personal hygiene, and danger of eating anywhere on-site except in the facilities provided and only after proper hand washing; risk of ingestion of contaminant through smoking with dirty hands; risks of smoking in gas-hazardous areas.

Training, which may take place at convenient times and locations on the site, should aim to get the essential message over to the person at the working face in a simple direct manner, without creating uncertainty through ambiguity. The instruction may be usefully supplemented by visual aids, posters, notices, area demarcations, etc. set out on-site.

23.5.2 Safety officer

It is recommended in Section 22 that the reclamation Contract requires the Contractor to provide a qualified safety officer. Whether this appointment is to be a full-time job or combined with other work will depend on the size of the site. However, the safety officer or a delegated and qualified representative must be present on site whenever work is in progress in hazardous areas.

Most contractors' safety officers are familiar with the normal hazards of civil and structural engineering. For derelict sites the safety officer should also have a grounding in chemistry and experience in gas monitoring and sampling, as well as sound knowledge of the risks of excavation in contaminated or unstable ground. It is probable that such composite background will not be found in one individual. Nevertheless, the required expertise must be available on-site, whenever hazardous work is undertaken.

23.5.3 Permits to work in restricted areas

No person should be allowed to enter a Restricted Area (Section 23.2.2) without a permit, issued each day by authority of the safety officer and surrendered at the end of the working shift.

Permits should not be issued until the safety officer is satisfied that:

1. Personnel have adequate protective clothing for the ambient conditions
2. All necessary safety equipment and monitoring equipment is to hand and functioning at the work place, as well as emergency rescue equipment (as required)
3. Sufficient, properly trained personnel will be present at the work place to deal with an emergency
4. Excavations have been safeguarded as Section 23.5.4 (unstable sides protected, atmosphere gas-free, pumps and tanks ready for dealing with liquids); and that
5. All emergency measures, outlined in Section 23.5.5 are currently in force.

In practical terms, permits may correspond to special colour coding of clothing (e.g. coloured safety helmets). Before entering Restricted Areas personnel should surrender matches and lighters.

23.5.4 Excavations and pits in restricted areas

During site-investigation works, when sophisticated safety measures are not usually called for, no-one should enter a trial pit. Samples should be taken from the contents of the digger bucket or, with long-handled tools, from the pit itself.

At other times during reclamation works, personnel should be warned of the risks of entering pits, trenches or other excavations. Before a person enters an excavation over 1.2 m deep for any purpose whatsoever in a Restricted Area, precautions must be taken as follows:

* Either the sides must be battered or supported, or, in emergency, a safety cage must be available for lowering into the excavation
* The atmosphere must be proved acceptably gas-free immediately prior to entry (see also Section 23.3.1 on gases)
* Where gas is present, gas detecting or monitoring equipment must be in position, to sound an audible warning of deteriorating conditions
* Safety equipment must be to hand for emergency operation and rescue, including ladders in strategic positions and audible alarms
* At least three persons must be present, *one of whom must remain on the surface.*

Gas-monitoring equipment should be located at or near the advancing face of an excavation. The toxic gases to be detected will depend on site history, but oxygen deficiency and flammable gas should always be monitored. Site personnel should also be encouraged to observe by sight or sense of smell any unusual conditions.

Where oxygen is deficient, normal respiratory face masks are ineffective. Positive-pressure breathing apparatus gives protection but is awkward to wear for long periods. Where conditions permit, a preferable alternative would be a purge-ventilation system at the work face to dilute toxic gases to below safety levels.

Particular care is needed in dealing with old pipes, drains, sumps, tanks and drums, unearthed during excavation, so as not to release dangerous liquids. (See also Section 23.6.2.) Special tanks should be available as necessary for the temporary storage of liquids.

Centrifugal pumps may not be suitable for handling the types of sediment, sludges and viscous liquids that may be encountered in excavations; positive displacement sludge-pumps of non-corroding material may be required.

23.5.5 Emergency procedures

Even with proper safety procedures in operation, unforeseen emergencies may arise, and plans for effective countermeasures should be in hand before work starts. Emergencies may be caused by unstable excavations, disturbance of dangerous liquids, gas hazards in excavations, collapse of underground voids, fire, explosion or similar hazards of derelict sites, where the consequences may be sufficiently great to require outside aid. Countermeasures will depend very much on the type of site, and only general guidance can be given here. Effective countermeasures will require not only correct tactical action by site workers but also the existence of effective strategic pre-planning.

Personnel should be trained to recognise unusual conditions, as well as to know the drill for sounding the alarm and actions to be taken on hearing it. *All alarm and safety equipment should be to hand alongside the working face. Speed will be essential in rescuing persons from excavations, but personnel must be warned against entering such areas alone and without safety equipment. Rescue drill must be an essential part of all workers' training.*

Before hazardous work starts, the relevant local emergency services should be alerted to the impending works, with details of possible hazards, countermeasures in emergency, means of access, names of key personnel, and similar information. Properly signposted vehicle access to the site (and to the working face) must be maintained uninterrupted, and a site telephone should be always available for emergency calls.

23.6 Containment of contamination

All works on-site must be so conducted as to ensure, at all times, secure containment of site contamination. Containment applies to all three phases: solid, liquid, and gaseous.

23.6.1 Solids

In all operations contamination should not be transferred to clean areas of the site. If contaminants are moved to other 'dirty' areas prior to site investigation, the local build-up of material must be taken into account in the subsequent sampling.

Any transfer of contaminated material to areas offsite must have the prior approval of the appropriate waste-disposal authority and, where applicable, the Health and Safety Executive. Materials classified as special wastes may only be deposited at tips specially licensed in accordance with the Control of Pollution Act 1974.

No contaminant should be dumped into watercourses which may carry the undesirable material beyond the confines of the site. Working faces should be sprayed with water as necessary to prevent dried-out dust from blowing away from the site. The quantity of water used should be controlled to avoid creating contaminated mud.

23.6.2 Liquids

During demolition care should be taken not to breach any tank, drum or other container without first checking that it contains no dangerous liquid. Special care is required not to breach the banks of settlement lagoons and tar lagoons. No construction work, excavation or backfilling of part-empty tanks, tar wells, etc. should start until all liquid has been evacuated.

Plant operators should be continually alert to the need for great care in perforating the floor of any underground tank or other void encountered in driving boreholes, piles, vibrated stone columns, etc., so as to prevent the release of contaminant into clean ground or the aquifer. When drilling through contaminated ground into a permeable horizon, special precautions (cement-bentonite sealing, etc.) are essential to prevent migration of contaminant into the aquifer.

Temporary tanks should be provided for the emergency storage of dangerous liquids, as well as sandbags, clay or similar materials for the bunding of liquids inadvertently released. No spilt liquid must be allowed to escape into a drain or other watercourse. No liquid should be pumped away from the site without the approval of the statutory authority.

Unsuitable containers for liquids can be dangerous. Proper containers should be provided and great care taken in transferring flammable liquids from one container to another; liquids should be pumped, not poured. Where site conditions permit fires, flammable liquids in small quantity are best disposed of by controlled burning, but never put on an already-burning fire. They must never be poured away down drains.

Flame cutting must be prohibited on vessels used for containing flammable liquids, unless they have been thoroughly purged of flammable gas or vapour. It is not enough to fill the drum with hot or cold water and then empty it.

23.6.3 Gases

Excavation may unavoidably release hazardous gases through drains or culverts, etc. onto adjacent lands. Prior consultation is essential with the local authorities and occupiers as well as careful monitoring during and after such operations.

23.6.4 Radioactivity

Radioactive material must not be removed from site without prior approval. Contaminated protective clothing and equipment must be safely contained on-site and disposed of as instructed by the National Radiological Protection Board or the Radiological Inspectorate of HSE.

23.6.5 Personnel and plant

On contaminated sites proper changing facilities must be provided to ensure that personnel do not take their working clothes, boots and tools off-site. Depending on site conditions, all personnel should shower and all vehicles and plant must be washed before they leave the site. Disposal of all wash water must conform with the requirements of the water authority and/or HSE. Where airborne dusts are present, the bodywork of vehicles should be cleaned.

23.7 Further information

Publications of HSE include advisory guides on many aspects of safety at work. Free issues, likely to be of interest in derelict land working, include specific leaflets from the following series:

- IND (G)
- Medical Series
- SS (Summary sheets for small contractors)
- EC (Environmental contaminants).

These are all obtainable from:

Health and Safety Executive
Library and Information Services
Baynards House
1 Chepstow Place
Westbourne Grove
London W2 4TF
Tel: 071-221 0870

who also issue periodically an updated publications list.

The following British Standards should be consulted for specific aspects of safety equipment:

BS 2091 *Specifications for respirators for protection against harmful dust and gases*
BS 4275 *Recommendations for the selection, use and maintenance of respiratory protective equipment*
BS 6408 *Specification for clothing made from coated fabrics for protection against wet weather.*

Barry (1985, 1987) has described safety precautions on contaminated sites.

References

ARBER, M.G. and VIVIAN, H.E. (1961)
Protection of mortars from sulphate attack by steam curing and carbonation.
Australian Journal of Applied Science, 1961, **12,** 330.

ARMSTRONG, G.D. (1983)
Local authority involvement in the decontamination of a tar distillery site.
London Environmental Supplement, Summer 1984, No. 7, 10–13.
London Scientific Services, Land Pollution Group.

ASSINK, J.W. (1985)
Extractive methods for soil decontamination; a general survey and review of operational treatment installations.
Proc. 1st Int. TNO Conf. on Contaminated Soil, Utrecht, Nov. 1985.
(Assink, J.W. and Van den Brink, W.J. ed.), 655–667.

BAKER, W. (1987)
Investigation Strategy.
Lecture at City of Birmingham Development Department Symposium on Methane Generating Sites, 9 Dec. 1987, Industrial Research Laboratories, Birmingham.

BALL, M.J. (1979)
The investigation of a quarry waste lagoon on the M.42 motorway.
Proc. Symp. on Engineering Behaviour of Industrial and Urban Fill, Birmingham. April 1979.
The Midland Geotechnical Society, C11–C24.

BARBER, C., MARIS, P.J. and KNOX, K. (1977)
Groundwater sampling: the extraction of interstitial water from cores of rock and sediments by high-speed centrifuge.
Water Research Centre Technical Report 54, 1977.

BARBER, C. (1983)
Treatability and treatment of leachate from domestic waste in landfills.
Proc. Int. Landfill Reclamation Conference: "Reclamation '83". 362–373.
Industrial Seminars, Tunbridge Wells.

BARRY, D.L. (1983)
Material durability in aggressive ground.
CIRIA Report 98, 1983.

BARRY, D.L. (1984)
Former iron and steelmaking plants.
In: *Contaminated Land—Reclamation and Treatment.* (Smith, M.A. ed.) 311–339.
Plenum Press, New York, 1985.

BARRY, D.L. (1987A)
Safety in site reclamation.
In: *Reclaiming Contaminated Land.* (Cairney, T. ed.) 181–199.
Blackie, London, 1987.

BARRY, D.L. (1987B)
Hazards from methane and carbon dioxide.
In: *Reclaiming Contaminated Land.* (Cairney, T. ed.) 223–255.
Blackie, London, 1987.

BECKETT, M.J. and SIMS, D.L. (1985)
Assessing contaminated land: UK policy and practice.
Proc. 1st Int. TNO Conf. on Contaminated Soil, Utrecht, Nov. 1985.
(Assink, J.W. and Van den Brink, W.J. ed.) 285–293.

BEEVER, P.F. (1982)
Understanding the problems of spontaneous combustion.
Fire Engineers' Journal, Sept. 1982, 38–39.

BEEVER, P.F. (1985)
Assessment of fire hazard in contaminated land.
Proc. 1st Int. TNO Conf. on Contaminated Soil, Utrecht, Nov. 1985.
(Assink, J.W. and Van den Brink, W.J. ed.) 515–522.

BRIDGES, E.M. (1987)
Surveying Derelict Land.
Clarendon Press, Oxford, 1987.

BRITISH STANDARDS INSTITUTION

BS1016:	Methods of analysis and testing of coal and coke (21 parts)
BS1047: 1983	Specification for air cooled blast furnace slag aggregate for use in construction
BS3882: 1965	Recommendations and classification for topsoil
BS4275: 1974	Recommendations for the selection, use and maintenance of respiratory protective equipment
BS5925: 1980	Code of practice for the design of buildings: ventilation principles and designing for natural ventilation
BS5930: 1981	Code of practice for site investigations
BS6031: 1981	Code of practice for earthworks
DD175: 1988	Code of practice for the identification of potentially contaminated land and its investigation
BS8110:	Structural use of concrete (3 parts)

BRE (Building Research Establishment) (1977)
Digest 206: *Ventilation requirements.*
Building Research Establishment, 1977.

BRE (1978)
Digest 210: *Principles of natural ventilation.*
Building Research Establishment, 1978.

BRE (1981)
Digest 250: *Concrete in sulphate-bearing soils and ground waters.*
Building Research Establishment, 1981.

BRE (1983A)
Digest 274: *Fill—classification and load carrying characteristics.*
Building Research Establishment, 1983.

BRE (1983B)
Digest 275: *Fill—site investigation, ground improvement and foundation design.*
Building Research Establishment, 1983.

BRE (1983C)
Digest 276: *Hardcore.*
Building Research Establishment, 1983.

BUSH, P.W. and COLLINS, W.G. (1972)
The application of aerial photography to surveys of derelict land in the United Kingdom.
Brit. Symp. on Remote Sensing, University of Bristol, Oct. 1972, 192.

CAIRNEY, T. (1984)
A rational approach to chemically contaminated land.
Proc. Inst. Civil Engrs. Part 2, Sept. 1984, Vol. 77, Tech. Note 422, 387–395.

CAIRNEY, T. (1985A)
Accelerated techniques for predicting movement of contaminants in soils.
EEC/DoE Research Contract. ENV/675/UK (H).

CAIRNEY, T. (1985B)
Soil cover reclamation experience in Britain.
Proc. 1st Int. TNO Conf. on Contaminated Soil, Utrecht, Nov. 1985.
(Assink, J.W. and Van den Brink ed.) 133–135.

CAIRNEY, T. (1987)
Soil cover reclamations.
In: *Reclaiming Contaminated Land* (Cairney, T. ed.) 144–169.
Blackie, London, 1987.

CARPENTER, R.J., GOAMAN, H.F., LOWE, G.W. and PECKSEN, G.N. (1985)
Guidelines for site investigation of contaminated land.
London Environmental Supplement, Summer 1985. No. 12.
London Scientific Services, Land Pollution Group.

CARPENTER, R.J. and LOWE, G.W. (1987)
Private communication.
London Scientific Services, Land Pollution Group.

CHARLES, J.A. (1973)
Correlation between laboratory behaviour of rockfill and field performance with particular reference to Scammonden Dam.
Ph.D. Thesis, University of London.

CHARLES, J.A., EARLE, E.W. and BURFORD, D. (1978)
Treatment and subsequent performance of cohesive fill left by opencast ironstone mining at Snatchill experimental housing site, Corby.
Proc. Conf. on Clay Fills, Inst. Civil Engrs., London 1978, 63–72.

CHARLES, J.A. (1979)
Field observations of a trial of dynamic consolidation on an old refuse tip in the east end of London.
Proc. Symp. on Engineering Behaviour of Industrial and Urban Fill, Birmingham.
April 1979. The Midland Geotechnical Society, E1–E13.

CHARLES, J.A. and DRISCOLL, R.M.C. (1981)
A simple in-situ load test for shallow fill.
Ground Engineering, Jan. 1981, **14** (1), 31–36.

CHARLES, J.A. and BURLAND, J.B. (1982)
Geotechnical considerations in the design of foundations for buildings on deep deposits of waste materials.
Structural Engineer 1982, **60A** (1), 8–14.

CHARLES, J.A., HUGHES, D.B. and BURFORD, D. (1984)
The effect of a rise of water table on the settlement of backfill at Horsley opencast coat mining site, 1973 to 1983.
In: *Ground Movements and Structures.* (Geddes, J.D. ed.) 423–442.
Pentech Press, London, 1985.

CHARLES, J. A. (1984)
Settlement of fill.
In: *Ground Movements and their Effects on Structures.* (Attewell, P.B. and Taylor, R.K. ed.) 26–45.
Surrey University Press, 1984.

CHENEY, J.E. (1973)
Techniques and equipment using a surveyor's level for accurate measurement of building movement.
Proc. Symp. on Field Instrumentation, British Geotechnical Society, London.
June 1973.
Butterworths, 1973, 85–100.

CHEYNEY, A.C. (1984)
Experience with the co-disposal of hazardous waste with domestic waste.
Proc. Symp. on Hazardous Waste Disposal and Re-use of Contaminated Land, London, July 1984.
Chemistry and Industry, 3rd September 1984, 609–615.

CHILDS, K.A. (1985)
—Management and treatment of groundwater: an introduction, 141–144.
—In-ground barriers and hydraulic measures, 145–182.
—Treatment of groundwater contaminated by leachate, 183–197.
—Mathematical modelling of pollutant transport by groundwater at contaminated sites, 199–205.
In: *Contaminated land—Reclamation and Treatment.* (Smith, M.A. ed.) 141–205.
Plenum Press, New York, 1985.

CIRIA (1982)
Building design legislation.
CIRIA Special Publication 23. 1982.

CIRIA (1985)
Scottish building legislation.
CIRIA Special Publication 34, 1985.

CLINE, J.F., GANO, K.A. and ROGERS, L.E. (1980)
Loose rock as biobarriers in shallow land burial.
Health Physics, Sept. 1980, Vol. 39, 497–504.

COULSON, M.G. and BRIDGES, E.M. (1985)
Site assessment and monitoring of contaminants by airborne multi-spectral scanner.
Proc. 1st Int. TNO Conf. on Contaminated Soil, Utrecht, 11th to 15th Nov. 1985.
(Assink, J.W. and Van den Brink, W.J. ed.) 365–378.

CROWHURST, D. (1987)
Measurement of gas emissions from contaminated land.
Building Research Establishment Report BR 100. 1987.

CROWHURST, D. and BEEVER, P.F. (1987)
Fire and explosion hazards associated with the re-development of contaminated land.
Building Research Establishment Information Paper IP. 2/87 April 1987.

CUR (1965)
Report 31: *Effect of various chemicals on concrete.*
Toevoegingen aan Beton Specie. Amsterdam, Commissie voor Uitvoering van Research 1965, 143.
Published as Translation 131 by Cement and Concrete Association, London.

DoE (Department of the Environment) (1978)
DoE Co-operative programme of research on the behaviour of hazardous wastes on landfill sites.
Final report of the Policy Review Committee.
HMSO, London, 1978.

DoE (1985a)
Circular 1/85: *The use of conditions in planning permissions.*
Dept. of the Environment, January 1985.

DoE (1985b)
The Building Regulations.
HMSO, London, 1985.

DoE (1986)
Waste Management Paper No. 26. *Landfilling wastes: a technical memorandum for the disposal of wastes on landfill sites.*
HMSO, London, 1986.

DoE (1987)
Circular 21/87: *Development of contaminated land.*
Dept. of the Environment, August 1987.

DoE (1989a)
A Review of Derelict Land Policy.
Dept. of the Environment, September 1989.

DoE (1989b)
Circular 17/89: *Landfill Sites: Development Control.*
Dept. of the Environment, 1987.

DEPARTMENT OF TRANSPORT (1986)
Specification for highway works.
HMSO, London, 1986.

DOWNIE, A.R. and TREHARNE, G. (1979)
Dynamic consolidation of refuse at Cwmbran.
Proc. Symp. on Engineering Behaviour of Industrial and Urban Fill, Birmingham.
April 1979. The Midland Geotechnical Society. E15–E24.

DUMBLETON, M.J. (1979)
Historical investigation of site use.
Proc. Conf. on Reclamation of Contaminated Land, Eastbourne, October 1979.
Society of Chemical Industry, London, 1980. B3/1–B3/13.

DUMBLETON, M.J. and WEST, G. (1970)
Air photograph interpretation for road engineers.
DoE: Transport and Road Research Laboratory Report LR 369.

DUMBLETON, M.J. and WEST, G. (1976)
Preliminary sources of information for site investigation in Britain.
DoE: Transport and Road Research Laboratory Report LR 403 (revised edn. 1976).

DURN, T.W.A. (1983)
Development problems of refuse tips in Hillingdon.
London Environmental Supplement, Summer 1984. No. 7, 24–25.
London Scientific Services, Land Pollution Group.

GREENWOOD, D.A. and KIRSCH, K. (1984)
Specialist ground treatment by vibratory and dynamic methods.
Proc. Conf. on Piling and Ground Treatment, Inst., Civil Engrs., London.
March 1983, Thomas Telford, 1984.

GREENWOOD, D.A. and THOMSON, G.H. (1984)
ICE Works Construction Guide. *Ground stabilization—deep compaction and grouting.*
Thomas Telford, London, 1984.

HARRIS, M.R.R. (1979)
Geotechnical characteristics of landfilled domestic refuse.
Proc. Symp. on Engineering Behaviour of Industrial and Urban Fill, Birmingham.
April 1979. The Midland Geotechnical Society. B1–B10.

HARRISON, W.H. (1987)
Durability of concrete in acidic soils and waters.
Concrete, Feb. 1987, 18–24.

HAWKINS, A.B. and PINCHES, G.M. (1987)
Cause and significance of heave at Llandough Hospital, Cardiff—a case history of ground
floor heave due to gypsum growth.
Quarterly Journal of Engng Geol., 1987, Vol 20,41–57.

HEALTH AND SAFETY EXECUTIVE (1981)
First-aid at work. HS(R)11.
HMSO, London, 1981.

HEALTH AND SAFETY EXECUTIVE (1981)
Guidance Note 40/89. *Occupational exposure limits.*
HMSO, London, 1989.

HMIP (Her Majesty's Inspectorate of Pollution) (1989)
Waste Management Paper No. 27. *The control of landfill gas: a technical memorandum on
the monitoring and control of landfill gas.*
HMSO, London, 1989.

HMSO (1988)
Problems arising from the redevelopment of gas works and similar sites (Second Edition).
HMSO, London, 1988.

HMSO (1990a)
House of Commons Environment Committee Report on Contaminated Land.
HMSO, London, 1990.

HMSO (1990b)
Development on unstable land.
DoE Planning Policy Guidance, PPG14.
HMSO, London, 1990

HILBERTS, B., EIKELBOOM, D.H., VERHEUL, J.H.A.M. and HEINIS, F.S. (1985)
In-situ techniques.
Proc. 1st Int. TNO Conf. on Contaminated Soil, Utrecht, Nov. 1985.
(Assink, J.W. and Van den Brink, W.J. ed.) 679–698.

HOLLAND, J.E. and LAWRANCE, C.E. (1980)
Behaviour and design of housing slabs on filling.
Proc. 3rd Australia/New Zealand Conf. on Geomechanics, Wellington, New Zealand,
May 1980. Vol. 1, 25–31.

HUCK, WALKER, and SHIMONDLE (1980)
*Innovative geotechnical approaches to the remedial in-situ treatment of hazardous materials
disposal sites.*
Proc. Nat. Conf. on Control of Hazardous Material Spills, Louisville, Kentucky, May 1980.

ICE (Institution of Civil Engineers) (1987)
Model Specification for Ground Treatment.
Instn. of Civil Engrs., London, 1987.

ICRCL (Interdepartmental committee for the reclamation of contaminated land).
—17/78. *Notes on the redevelopment of landfill sites.*
  7th edn. May 1988.
—18/79. *Notes on the redevelopment of gasworks sites.*
  5th edn. May 1986.
—23/79. *Notes on sewage works and farms.*
  2nd edn. November 1983.
—42/80. *Notes on scrapyards and similar sites.*
  2nd edn. October 1983.
—59/83. *Guidance on assessment and redevelopment of contaminated land.*
  2nd edn. May 1983.
—61/84. *Notes on fire hazards of contaminated land.*
  2nd edn. July 1986.
—64/85. *Asbestos on contaminated sites.*
  1st edn. May 1985.
Dept. of the Environment, Central Directorate on Environmental Pollution, London.

JARRETT, P.J.D. (1979)
Control and treatment of leachate from industrial and domestic wastes.
Gas emission from waste landfill sites—its regulation and control.
Proc. Symp. on Engineering Behaviour of Industrial and Urban Fill, Birmingham.
April 1979. The Midland Geotechnical Society. B17–B34.

KHAN, A.Q. (1979)
Investigation and treatment of Ravenfield tip.
Proc. Conf. on Reclamation of Contaminated Land ("Reclan"), Eastbourne, October 1979.
Soc. of Chem. Ind., London. F3/1–F3/11.

KILKENNY, W.M. (1968)
A study of the settlement of restored opencast coal sites and their suitability for building development.
Dept. of Civil Engng., University of Newcastle-upon-Tyne.
Bulletin No. 38.

KNIPE, C. (1979a)
Comparison of settlement rates on backfilled opencast mining sites.
Proc. Symp. on Engineering Behaviour of Industrial and Urban Fill, Birmingham.
April 1979. The Midland Geotechnical Society. E81–E98.

KNIPE, C. (1979b)
Contribution to discussion on "spontaneous heating".
Proc. Symp. on Engineering Behaviour of Industrial and Urban Fill, Birmingham.
April 1979. The Midland Geotechnical Society. E152.

de KREUK, J.F. (1985)
The microbiological decontamination of excavated soil.
Proc. 1st Int. TNO Conf. on Contaminated Soil, Utrecht, No. 1985.
(Assink, J.W. and Van den Brink, W.J. ed.) 669–678.

KRIZEK, R.J. and SALEM, A.M. (1977)
Field performance of a dredgings disposal area.
Proc. Conf. on Geotechnical Practice for Disposal of Solid Waste Materials, University of Michigan, ASCE 1977, 358–383.

de LEER, E.W.B. (1985)
Thermal methods developed in the Netherlands for the cleaning of contaminated soil.
1st Int. TNO Conf. on Contaminated Soil, Utrecht, Nov. 1985.
(Assink, J.W. and Van den Brink, W.J. ed.) 645–654.

LEIGH, W.J.P. and RAINBOW, K.R. (1979)
Observations of the settlement of restored opencast mine sites.
Proc. Symp. on Engineering Behaviour of Industrial and Urban Fill, Birmingham.
April 1979. The Midland Geotechnical Society. E99–E128.

LINDFORS, L.G. (1979)
Reclamation of the site of a herbicide factory.
Proc. Conf. on Reclamation of Contaminated Land ("Reclan"), Eastbourne, October 1979.
Society of Chemical Industry, London, 1980. F8/1–F8.4.

LOWE, G.W. (1984a)
Dealing with polluted sites.
Building Surveyors' Weekend Briefing, March 1984.
University of Warwick. 385/0958W/RA(1).

LOWE, G.W. (1984b)
Investigation of land at Thamesmead and assessment of remedial measures to bring contaminated sites into beneficial use.
Proc. Nat. Conf. and Exhibition on Management of Uncontrolled Hazardous Waste Sites, Washington D.C., November 1984.

LYONS, V.R.M. (1984)
Investigation of contaminated land in Southwark.
London Environmental Supplement, Summer 1984. No. 7, 17–21.
London Scientific Services, Land Pollution Group.

MARSLAND, A. and QUARTERMAIN, R. (1974)
Further development of multipoint magnetic extensometers for use in highly compressible ground.
Geotechnique 1974. Technical Note 429.

McCARTHY, M.J. (1979)
Reclamation of a refuse tip for open space and housing development.
Proc. Conf. on Reclamation of Contaminated Land ("Reclan"), Eastbourne, October 1979.
Society of Chemical Industry, London, 1980. B8/1–B8/11.

MEYERHOF, G.G. (1951)
Building on fill with special reference to the settlement of a large factory.
Structural Engineer, November 1951, Vol. **XXIX** (No.2). 46–57.

NHBC (National House-Building Council) (1985)
Practice Note 3: *Building near trees.*
National House-Building Council, 1985.

NRPB (National Radiological Protection Board) (1986)
Radiological significance of the utilization and disposal of coal ash from power stations.
NRPB Contract 7910–1462. Report, January 1986.
NRPB., Didcot, Oxfordshire.

NAYLOR, J.A., ROWLAND, C.D., YOUNG, C.P. and BARBER, C. (1978)
The investigation of landfill sites.
Water Research Centre. Technical Report 91, October 1978.

NETHERTON, D.W. and TOLLIN, B.I. (1982)
Reclamation of the Beckton Alps and adjacent areas.
Symp. on Development of Contaminated Land, London, May 1982.
The Public Health Engineer, October 1982, Vol. 10. No. 4, 202–211.

NIXON, P.J. (1978)
Floor heave in buildings due to the use of pyritic shales as fill material.
Chemistry and Industry, 4th March 1978, 160–164.

OECD (Organisation for Economic Co-operation and Development) (1979)
Design Manual. Road Research: Construction of roads on compressible soils. (77 80 02 1).
Organisation for Economic Co-operation and Development, Paris, 1979.

PADFIELD, C.J. and SHARROCK, M.J. (1983)
Settlement of structures on clay soils.
CIRIA Special Publication No. 27.

PARKER, A., PEARCE, K.W., FENNER, S.F. and WRIGHT, M.S. (1984)
The decontamination of a disused gasworks site.
Journal of Hazardous Materials **9,** 347–354.
Elsevier, Amsterdam, 1984.

PARRY, G.D.R. and BELL, R.M. (1985)
Covering systems.
In: *Contaminated Land—Reclamation and Treatment.* (Smith, M.A. ed.) 113–139.
Plenum Press, New York, 1985.

PECKSEN, G.N. (1985)
Methane and the development of derelict land.
London Environment Supplement, Summer 1985. No. 13.
London Scientific Services, Land Pollution Group.

PULFORD, I.D. and DUNCAN, H.J. (1978)
Predicting the potential acidity in reclaimed coal mine waste.
Reclamation Revue 1978. Vol. 1, 36–37.

ROBERTS, T.M. and GEMMELL, R.P. (1979)
Establishment of vegetation on gasworks waste.
Proc. Conf. on Reclamation of Contaminated Land, Eastbourne ("Reclan"), October 1979.
Society of Chemical Industry, London, 1980. F11/1–F11/10.

RULKENS, W.H., ASSINK, J.W. and van GEMERT, W.J.Th. (1984)
On-site processing of contaminated soil.
In: *Contaminated Land—Reclamation and Treatment.* (Smith, M.A. ed.) 37–90.
Plenum Press, New York, 1985.

RYS, L.J. and JOHNS, A.F. (1985)
The investigation and development of a landfill site.
Proc. 1st Int. TNO Conf. on Contaminated Soil, Utrecht, November 1985.
(Assink, J.W. and Van den Brink, W.J. ed.) 625–636.

SANNING, D.E. (1985)
In-situ treatment.
In: *Contaminated Land—Reclamation and Treatment.* (Smith, M.A. ed.) 91–111.
Plenum Press, New York, 1985.

SHERWOOD, P.T. (1987)
ICE Works Construction Guide. *Wastes for imported fill.*
Thomas Telford, London, 1987.

SMITH, M.A. (1985)
Contaminated Land—Reclamation and Treatment. (Smith, M.A. ed.)
Plenum Press, New York, 1985.

SMITH, M.A. and BELL, R.M. (1985)
Upward movements of metals into soil covering metalliferous waste.
Proc. 1st Int. TNO Conf. on Contaminated Soil, Utrecht, Nov. 1985.
(Assink, J.W. and Van den Brink, W.J. ed.) 133–135.

SMITH, M.A. and ELLIS, A.C. (1986)
An investigation into methods used to assess gasworks for reclamation.
Reclamation and Revegetation Research. 1986. Vol. 4. 183–209.

SMITH, M.A. (1987a)
Fly-ash disposal and ultilization: environmental considerations.
Proc. Int. Symp. on Ash: a Valuable Material, CSIR Conference Centre, Pretoria,
South Africa, February 1987.

SMITH, M.A. (1987b)
Demonstration of remedial actions and technology for contaminated land and groundwater.
Summary presented at Int. Studies on Technology for Cleaning Contaminated Land and
Groundwater.
Proc. Land Reclamation 88. Durham County Council 1988.

SMITH, M.A. (1990)
Data analysis and interpretation.
In: *Recycling Derelict Land.* (Fleming G. ed.)
Thomas Telford Ltd. (London, 1990).

SMYTH-OSBOURNE, K.R. and MIZON, D.J. (1984)
Settlement of a factory on opencast backfill.
In: *Ground Movements and Structures.* Vol. 3 (ed. Geddes, J.D.) 463–479.
Pentech Press, London, 1985.

SOMOGYI, F. and GRAY, D.H. (1977)
Engineering properties affecting disposal of red muds.
Proc. Conf. on Geotechnical Practice for Disposal of Solid Waste Materials, University of Michigan. ASCE, 1977.

SOWERS, G.F. (1973)
Settlement of waste disposal fills.
Proc. 8th Int. Conf. on Soil Mechanics and Foundation Engineering, Moscow, 1973. Vol. 2.2. 207–210.

STEWARD, H.E. and CRIPPS, J.C. (1983)
Engineering implications of weathering of pyritic shales.
Quarterly Journal of Engng. Geol. 1983, Vol. **16,** 281–289.

SYMONS, I.F. and MURRAY, R.T. (1988)
Conventional retaining walls: pilot and full scale studies.
Proc. Inst. Civil Engrs. Part 1. June 1988. Vol. 84. Paper 9306, 519–538.

TAYLOR, R.K. and CRIPPS, J. C. (1984)
Mineralogical controls on volume change.
In: *Ground Movements and their Effects on Structures.* (Attewell, P.B. and Taylor, R.K. ed.) 268–303.
Surrey University Press, 1984.

THOMAS, G.H. (1983)
Properties of iron and steel slags.
Proc. Conf. on Reclamation of Former Iron and Steelworks Sites, Windermere, October 1983.
Durham County Council/Cumbria County Council. C1–C8.

THOMSON, G. and ALDRIDGE, J.A. (1983)
London docklands—problems associated with their development with special reference to the Surrey Docks scheme.
Proc. Int. Land Reclamation Conf. "Reclamation 83", 124–132.
Industrial Seminars, Tunbridge Wells, 1983.

THORBURN, S. and BUCHANAN, N.W. (1983)
Re-development of urban areas in Glasgow and Edinburgh.
Proc. Int. Land Reclamation Conf. "Reclamation 83", 133–138.
Industrial Seminars, Tunbridge Wells, 1983.

THORNTON, I. (1985)
Metal contamination of soils in UK urban gardens: implications to health.
Proc. 1st Int. TNO Conf. on Contaminated Soil, Utrecht, Nov. 1985.
(Assink, J.W. and Van den Brink, W.J. ed.) 203–209.

THREADGOLD, L. (1979)
Contribution to discussion.
Proc. Symp. on Engineering Behaviour of Industrial and Urban Fill, Birmingham. April 1979. The Midland Geotechnical Society. B79.

TOMLINSON, M.J. (1963)
Foundation Design and Construction.
Pitman Press, Bath. 1st edition, 1963.

TOMLINSON, M.J., DRISCOLL, R.M.C. and BURLAND, J.B. (1978)
Foundations for low-rise buildings.
The *Structural Engineer*, Part A, 1978, **56A** (6), 161–173.

WEATHERLEY, N. (1979)
Trench filled pfa in colliery waste supporting old peoples' bungalows.
Proc. Symp. on Engineering Behaviour of Industrial and Urban Fill, Birmingham. April 1979. The Midland Geotechnical Society. D71–D76.

WELSH OFFICE (1988)
Survey of contaminated land in Wales. (Revised edn. August 1988)
Welsh Office: Environmental Advisory Unit, 1988.

WELTMAN, A.J. and HEAD, J.M. (1983)
Site investigations manual.
CIRIA Special Publication 25, 1983.

WILDE, P.M. and CROOKE, J.M. (1979)
Problems and solutions in developing large areas of filled ground at Warrington New Town.
Proc. Symp. on Engineering Behaviour of Industrial and Urban Fill, Birmingham.
April 1979. The Midland Geotechnical Society. D39–D47.

WOOD, A.A. and ROSS, A.H. (1979)
Reclamation of agricultural land used for disposal of sewage and sewage sludges.
Proc. Conf. on Reclamation of Contaminated Land ("Reclan"), Eastbourne, October 1979.
Society of Chemical Industry, London, 1980. D8/1–D8/11.

WRc (1988a)
Pipeline installation in contaminated land.
WRc Report ER319E, 1988.

WRc (1988b)
Pipe materials selection manual.
WRc, Swindon, 1988.

WRc (1990)
Effects of soil contaminants on materials used for distribution of water.
WRc Report EC1 9168, Swindon, 1990.

YOUNG, C.P. and BARBER, C. (1979)
Contaminated land—effects on water resources.
Proc. Conf. on Reclamation of Contaminated Land ("Reclan"), Eastbourne, October 1979.
Society of Chemical Industry, London, 1980. D5/1–D5/11.

Appendix 1 Financial assistance for development

At present, market forces control the rate of development of derelict land. Derelict land is often cheap to buy, can attract land-reclamation grants and can be profitable if reclaimed at reasonable cost. The local authority is directly involved as public safeguard and, in many cases, as instigator of the development.

A1.1 DoE Derelict Land Grant

Circular 28/85 (DoE, 1985) sets out the minor revisions proposed for 1986/87 on Derelict Land Grants (DLG) for eligible sites covering costs of site investigation and, separately, costs of land improvement up to greenfield condition.

Current political conditions favour commercial development and are biased against state controls. However, the system outlined in Circular 28/85 still favours land improvement by the local authority, with the developer subsequently taking over the improved land. The DoE allows compensation for the improvement costs incurred (including site investigation) thus:

Land improved by	Assisted areas	Other areas
Local authority	100%	50%
Private developer	80%	50%

Despite an increasing flow of applications for grants from private developers (mainly for relatively small projects), grants to private developers accounted for only 10% of DoE funding in 1985.

A1.2 Eligibility for DLG

The grants available to the local authority and the developer in Assisted Areas are as follows. The local authority:

- Can have agreement on and obtain grant for 100% of costs up to one year in advance (three years in the special case of a rolling programme)
- Can get a 100% grant for site investigation, even if no development ensues
- Must pay back DLG, for land improvement only, if no development follows within ·three years.

On the other hand, the private developer:

- Can recover 80% of site investigation costs only after completion and if development follows
- Can recover, after completion, 80% of net costs of land improvement to greenfield status (net after deduction of gain derived from enhanced value).

The local authority thus has major incentives for large-scale derelict land improvement.

A1.3 Current procedure on land improvement

The procedure now commonly operating for large-scale developments under DLG in eligible Assisted Areas, where the local authority does not already own the derelict land, is as follows:

1. Local authority draws up master plan and discusses possibilities with the developers in each area, who may or may not be the landowner(s).
2. Local authority, with developer, submits plan to DoE with estimate of cost of land improvement.
3. Local authority applies to DoE for grant to cover net cost of land improvement, i.e.: purchase price of derelict land plus land improvement costs, less resale value of improved land (purchase price and resale value fixed by District Valuer).

4. Once the DLG is approved by DoE, the local authority buys the derelict land and carries out the site investigation and/or improvement to greenfield status, as the case may be.
5. The local authority sells the greenfield to the developer.
6. Developer must start project within three years.

A1.4 Private sector financing

An engineer advising a private client can recommend either the above procedures or choose from the other options outlined below. Where a private developer is also to carry out the land improvement, the works will have to be to the standard specified by the local authority.

DLG for private developers. The government has power to give full DLG to bodies other than local authorities, such as Groundwork Trusts, and wishes to encourage private owners to reclaim sites for immediate development.

Private sector investment. Financing through DLG suffers from limited resources; about £76 million was made available in 1985/86. This often means that the scheme is not attractive to the private sector on account of the time lag between application and grant. A case in point is the Stockley Park regeneration of a 150-ha waste tip for industrial and recreational development. Faced with a two-year waiting period for DLG, the development has proceeded by purely private-sector investment. DLG has, in fact, proved less interesting to private developers in south-east England, where development of inner-city dereliction can be very profitable without recourse to outside involvement.

Urban Development Grant (UDG). As an incentive to private development in other areas, where inner-city regeneration is urgently needed although less potentially profitable, there exists also the scheme of Urban Development Grants. These grants by central government are intended to offset, at least partially, the shortfall between (1) the actual value of a completed inner-city regeneration project, and (2) the return that a private investor might reasonably expect elsewhere. Eligibility for these grants requires substantial private sector investment. Up to 1987 about £65 million had been made available for use on projects involving over £290 million of private sector financing.

A1.5 Further information

Further information. DoE Circulars 28/85, 15/85 and 14/84, obtainable from HMSO, give further details of the government's intentions for grants. Enquiries for specific details as well as applications for grants should, in the first instance, be addressed to the relevant DoE regional office or to the Inner Cities Directorate of the DoE.

Appendix 2 Application to Scotland

The main text of this publication refers to derelict land reclamation and development as practised in England and Wales. The following notes, kindly supplied by the Industry Department of Scotland (IDS), outline the differences to be encountered in these same operations in Scotland. They were of valid application in August 1987.

The Scottish Development Agency has the statutory responsibility for the reclamation of derelict land in Scotland. The Agency's statutory powers are set out in Section 8 of the Scottish Development Agency Act 1975. This provides that, where necessary, it may acquire derelict and, as appropriate, adjacent land either by agreement or compulsorily for the purpose of bringing it into use or improving its appearance. The Agency does not have a general vesting power which would allow publicly owned derelict land to pass into its possession.

When the Agency was founded in 1975, the then-existing system for government grant aid to local authorities for land rehabilitation was abolished and direct budgetary provision made to the Agency. Local authorities can still undertake land rehabilitation but their budgets do not contain any specific allocations for funding them. A number of authorities have, however, carried out schemes as agents of the SDA. The Agency also sometimes makes a financial contribution to assist private developers clear derelict land, but it does not operate a grant scheme for private sector developers such as is now offered in England.

Local authorities normally place great emphasis upon derelict land in Structure and Local Plans and the positive development opportunities that exists. This is particularly strongly stressed in the Glasgow context, where the Glasgow Eastern Area Renewal (GEAR) initiative has been successful in regenerating inner urban investment. In this situation the procedure of Section A1.3 in Appendix 1 is not readily applicable to Scotland.

The relevant legislation applicable in Scotland is as follows:

Town and Country Planning (Structure and Local Plans) (Scotland) Regulations 1983 (SI 1983/1590)
Control of Pollution Act 1974 (c40)*
Rivers (Prevention of Pollution) (Scotland) Acts 1951 and 1965 (c66 and c13)
Water (Scotland) Act 1980 (c45)
Refuse Disposal (Amenity) Act 1978 (c3)

It should be noted that Regulations made under the Control of Pollution Act 1974 for Scotland may, on occasion, differ from those which apply in England and Wales.

The position under the Building Standards (Scotland) Regulations 1981 to 1986 is that there are no specific requirements for building on derelict land as such. There are, however, requirements for site preparation in Part 5 of the Regulations which, in broad terms, require the land to be drained and for any material harmful to either the building or its occupants to be removed. These requirements are similar to those of Part C of the Building Regulations 1985. Approved Document C does, however, go into rather more detail then the current Explanatory Memorandum on the Building Standards Regulations.

* UK legislation

Appendix 3   Basic principles applied to specific types of site

This appendix illustrates the application of the basic principles, established in Sections 11 to 20, to the particular requirements of the more common types of site likely to be encountered:

- Industrial sites (represented by gasworks)
- Commercial sites (represented by scrapyards)
- Domestic sites
- Opencast fill sites.

A3.1 Industrial sites

A3.1.1 General

Table A1 lists a number of types of industrial site and the associated hazards. Where the past history is known this table may assist in narrowing the range of contaminants to be sought in a given case.

Table A1 Industrial works and associated hazards

Industrial works	Associated hazards
Gasworks sites	See Section A3.1
Steelworks including slag dumps	Difficult to bore site investigation holes. Material expands on weathering and can adversely affect foundations. Possible presence of cyanide from heat treatment process. Corrosive conditions may exist. Special refractory bricks can contain contaminants. Coking plant areas have same hazards as gasworks
Smelting works	Heavy metal contamination including surrounding area. Low pH values
Refineries	Various including oil, tarry products, phenols
Chemical and explosive works	Heavy metals and other toxins. Highly variable (treat on own merits)
Tanneries	Sodium chromate, organic matter, anthrax spores (sterilise with formaldehyde)
Electricity-generating stations	Heavy foundations. Combustion from carbonaceous material if coal-fired. Timber from cooling towers can produce arsenic contamination if burned on site. Asbestos and ash deposits
Water treatment works	Sludges may contain heavy metals
Sewage-treatment works	Heavy metal contamination plus possible pathogens
Breweries	Effluents have high BOD
Glass-making works	Possibility of chemicals for colouring glass. Lead in crystal glass
Mineral-processing works: brick, cement, lime	Physical problems and possible high alkalinity
Pithead works	Heavy foundations. Buried coal if steam-driven at one time. Presence of electrolytes from current controllers if electrical winding
Metal-working factories	Oil/water emulsions from turnings can degrade to phenols. Oil contamination. Chemical pickling wastes such as cyanide and chromium, acids

Conversely, a specific site contaminant may be recognised which could indicate the type of industry that might have produced it. For these circumstances, Table A2 relates common pollutants to possible processes or industries.

This Section A3.1 outlines the procedures for site investigation, appraisal and treatment for the specific case of a gasworks site, as representative of industrial sites generally. Other types of industrial site would be dealt with in broadly the same way but, in each case, the relevant contaminants and potential hazards should be anticipated and the procedures planned accordingly.

Table A2 Possible sources of some commonly occurring contaminants (after DoE, 1978)

Contaminants	Possible sources
Heat treatment cyanide wastes	Principally from the heat treatment and carburising of metals
Metal finishing wastes	Cleaning, pickling, anodising, phosphating, rust-proofing, etching, electromachining, electroplating and galvanising of ferrous and non-ferrous metals
PCBs and analogue compounds	Mainly from uses in electrical equipment, but also from heat transfer and hydraulic fluids
Mineral oils	From lubrication, cutting and preserving oils
Solvents	A wide range of industries, including petrochemical, metal finishing, metal extraction, printing
Halogenated hydrocarbon solvents	Dry cleaning, metal cleaning, paint and printing industries
Mercury wastes including organo-mercury compounds	Primary cell manufacture, electrical equipment, agricultural chemicals, catalysts, chloralkali industry, metal refining, pharmaceuticals, anti-fouling paints and dentistry
Arsenical wastes	Extraction and smelting of metals, wood preservation, treatment of furs and hides, agricultural preparations, glass-making (good-quality crystal)
Halogenated organic wastes	Chemical industry
Tarry wastes including still bottoms	Gasworks, coal tar processing, oil refining, petrochemical, plastics, rubber and pharmaceutical manufacture, solvent and oil recovery
Asbestos wastes	Friction materials, thermal insulants, textile articles, building and construction materials
Plastic wastes including monomers and precursors	Manufacture of basic plastics, manufacture and use of plastic articles, paints, varnishes, adhesives and thermal insulants
Phenols, cresols and simple derivatives	Coke ovens, coal tar, oil refining and petrochemicals, plastic resins, varnishes, adhesives, dyestuffs and pharmaceutical products
Other organic compounds	Wide range of sources
Metals and their compounds including organometallics	Gasworks, mining, metal extraction and refining, production and use of metal compounds, water and sewage treatment
Acid wastes	Surface treatment, heavy chemical industry
Pharmaceutical wastes	Manufacture and use in medicine and agriculture
Medical waste and infected material	Medical and veterinary services, agriculture
Biocides, herbicides	Pest and plant control, paper and paint industries, wood textile and cooling-water treatment

A3.1.2 Gasworks sites

Coal carbonisation started in the UK in the eighteenth century and ceased in the 1970s (later in Northern Ireland) when rendered obsolete by the exploitation of natural gas. Over about a century and a half there were some 2000 gasworks in operation, ranging in area from less than one ha to 200 ha. Continual build-up of contamination resulted from carbonisation by-products and substances imported for purification and other processes. Much of the contamination is toxic and occurs in liquid or solid form. On many sites the precise location of specific buried pollutants today may be unknown.

Early reclamation of gasworks was not always well controlled, and widespread pollution has occurred on some sites by indiscriminate breaching or filling of tanks or regrading of dumps. The engineer should always act as if the contamination were still confined to specific parts of the site and take stringent precautions to prevent any escape of contaminant through careless working.

Materials such as spent oxide have, on occasion, been used as fill, thereby contaminating clean soil. Migration of water-soluble compounds such as phenols may be found to extend beyond the immediate confines of the works.

A3.1.3 Coal gas production process

Distillation of coal in fireclay retorts was used to produce 'town gas' consisting principally of hydrogen (48%) and methane (32%) but with ethylene, carbon monoxide, carbon dioxide and nitrogen. A residue of coke remained in the retort, while the liquid products, carried over as vapour with the hot gasses, included aqueous ammoniacal liquor and coal tar. The

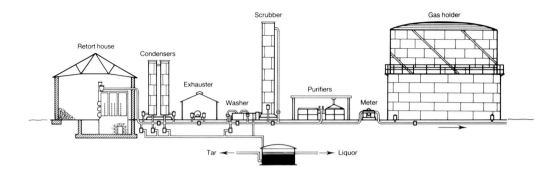

Figure A1 *Coal gas production process*

hot gases, taken off by a rising main, were freed of tar in a trap in the tar main and then passed into the gas main. Air-cooled condensers (Figure A1) extracted a further quantity of tar and ammonia. Gas flow was promoted by the exhauster which served to maintain a reduced pressure on the retorts. Washers and scrubbers extracted the remaining ammonia from the gas, leaving hydrogen sulphide and prussic acid as the significant impurities.

The latter were removed by passing the gas over trays of moist hydrated ferric oxide in the purifiers:

$$Fe_2O_3 + 3H_2S = Fe_2S_3 + 3H_2O$$
$$Fe_2O_3 + 6HCN = 2Fe(CN)_3 + 3H_2O$$

The iron oxide was revived from time to time by exposure of the sulphide to air:

$$2Fe_2S_3 + 3O_2 = 2Fe_2O_3 + 6S$$

and was re-used until it contained about 50% of sulphur. The 'spent oxide' (usually coloured blue by toxic cyanogen compounds) was stocked in dumps, used as fill on-site, or sold for conversion into sulphuric acid. The colloquial term for the waste was 'blue billy'.

A3.1.4 Gasworks: sources of contamination

Contamination may occur in various forms including:

Spent oxide: may be only partially spent. It is often of very low pH and contaminated with iron sulphide, free sulphur, cyanides and ammonium compounds and found in dumps or mixed with other surface materials.
Liquid residues: aromatics, coal tars, phenols, ammonium compounds contained in wells, tanks, retorts or pipelines or deliberately dispersed into the ground.
Miscellaneous sources: spillages, leakages, buried asbestos, buried pipework and tanks.

Pollution may occur from latent sources by the upward migration of contaminants dissolved in oil which may be squeezed to the surface under load, for example, by site traffic. Other undetected contaminants left untreated could migrate through soils and strata boundaries to spread pollution via the groundwater or by soil suction, or corrode buried construction materials.

A3.1.5 Gasworks: initial assessment

Contamination on a complex industrial site can only be assessed in approximate terms. Dangerous toxicants concentrated in small quantities may well be missed, unless the resolution of the sampling grid is made uneconomically fine. Consequently, much deductive work is necessary during the desk study to highlight obvious sources of contamination. In face of the uncertainty as to how representative any sampling may be, the site should be appraised on the basis of the maximum concentrations of contaminant discovered.

For a gasworks site, locations of the following structures and areas should be identified from the desk study:

- Coal stores
- Retort house
- Coke conveyors and stores
- Tar separators, tanks and wells
- Sulphate plant and stores
- Naphthalene and ammonia scrubbers
- Gas purifiers
- Compressors and governors
- Gas holders
- Pipe runs (above and below ground)
- Spent oxide dumps.

Gasworks residues which may contain some of the 'listed' contaminants of Table 15 include:

- Ammoniacal liquor
- Coal tars
- Naphthalenes, etc.
- Blue billy (spent oxide, sulphur, cyanide)
- Phenols, cresols.

As necessary, site inspection with trial holes should be made to confirm the siting of the items.

A3.1.6 Gasworks: site investigation

Gasworks sites are usually fairly large, and a first-stage investigation of the general levels of contamination should be carried out on a square or rectangular grid of between 25 and 50 m. A more closely spaced grid should be used at places of likely contamination identified in the initial assessment. Trial pits are superior to boreholes at this stage in identifying buried structures, and give a better indication of the type and extent of any obvious contamination.

Samples of the contaminant (if obviously visible) and of the surrounding soil should be collected for analysis under the supervision of the chemist in the team. A generous number of samples should be taken, say, at 0.5-m depth intervals. Not all need be analysed initially. Boreholes have a use in 'blind' sampling of the soil, and for providing information on the contamination at depth which may have been caused by leakage from spent oxide dumps or leakage of the more soluble contaminants such as phenol.

A generous number of samples of the surface materials should be taken and contaminants mapped. A gasworks may require a multi-stage site investigation with further sampling at a closer spacing in areas where the first sampling reveals high concentrations. Economic justification for a grid spacing of less than 25 m is doubtful (Smith and Ellis, 1986).

A3.1.7 Gasworks: basic chemical analyses

Only some of the samples taken should be analysed initially, the remainder being kept for further analysis if later found necessary. The initial analysis should be restricted to identifying those 'listed' contaminants (Table 15) likely to prejudice the proposed end use of the site. The advice of the chemist in the team should always be followed in determining the analytical programme.

For all but the least sensitive end uses (e.g. car parks) the initial analysis should seek to identify coal tars; PAHs; phenols; sulphur; sulphates; chlorides and free and complex cyanides. Values of pH should be measured.

For all types of end use, without distinction, the analysis must also check the known or suspected presence of asbestos and potential combustibility.

Gasworks sites are not normally considered for sensitive end uses where phytotoxicity or uptake of contaminants by food-crops may be a critical factor.

If such sensitive end use is contemplated, analysis for the following metals will be necessary:

Arsenic	Copper*
Cadmium*	Nickel*
Lead	Zinc*

For those metals marked with an asterisk, 'plant available' values should be determined.

A3.1.8 Gasworks: additional chemical analysis

In exceptional cases it may be necessary to identify contaminants beyond those 'listed' items on Table 15. Table A3 sets out possible additional contaminant hazards that may require checking. However, advice should always be sought from the team chemist before any such additional analyses are contemplated.

Table A3 Additional contaminant hazards encountered on gasworks sites

Metals		
	Antimony	Mercury
	Barium	Molybdenum
	Cobalt	Vanadium
	Chromium (3 and 6-valent)	
	Iron	
Anions		
	Nitrate	Phosphate
Organic compounds		
	Cyanate	Thiocyanate
	Ferricyanide	Organic sulphides
	Ferrocyanide	
Various		
	COD (chemical oxygen demand)	

Note: These items are additional to the basic listed hazards set out in Table 15. Not all will be important for a given end use, and for several of them ICRCL 'trigger values' are not yet available. A programme of analyses should take account of the possible sensitivity of a given target to a particular contaminant, and expert scientific advice will be needed (see also Section A3.1.7).

A3.1.9 Gasworks: groundwater

The groundwater regime at a site of this type is important, and rigorous investigation is advisable to establish flow patterns, concentrations of dissolved pollutants and any trends with time. A generous number of piezometers at various horizons is required, together with water sampling. If it is feared that contamination is spreading into an aquifer, observation wells may be necessary outside the area of the site. The hydrogeology of the site should be determined after demolition, to take account of any change that this work may produce. A monitoring period of at least 8 months is ideally required to record fluctuations in groundwater level. Failing this, readings should be taken during March and April when the groundwater table is usually at a maximum in the UK. Precautions to be observed when drilling through contaminated land in the vicinity of aquifers are described in Sections 15.5.3 and 15.5.4.

A3.1.10 Gasworks: appraisal

It is not possible to define acceptable or unacceptable concentrations of toxic substances for a given site. Each site has to be judged on the basis of all data available and the advice of the chemist should be sought. The problems likely to be posed by concentrations as measured will depend on the effectiveness of practicable remedial measures and possible effects of groundwater movement.

Careful consideration should be given to the end use planned for the site and the exposure routes for those likely to be at risk. As an illustration, concentrations above 3 mg/kg of cadmium may be undesirable if the land is intended for allotments, gardens or agricultural uses.

The appraisal should be made in accordance with the principles set out in Section 17. In the specific case of a gasworks site, much useful guidance is given in ICRCL Note 18/79, 5th edition (1986).

A3.1.11 Gasworks: remedial treatment

In the UK remedial treatment usually involves the capping of gasworks contamination with a cover layer designed according to the principles set out in Section 19.1. Fresh tar spills, however, can so degrade the mechanical properties of a soil that complete excavation and removal may become the only recourse. The engineer should therefore plan and specify remedial works with extreme care. Where tar wells, etc. are intact, they should be emptied of tarry materials and water, completely filled with rubble and capped. Tars and oils are more easily handled if mixed with sand, pfa, or, better, granular bentonitic materials.

Even with covering layers, some excavation and removal of contaminated ground, especially spent oxide, may be necessary. Such contamination will almost certainly be classified as 'special waste' and may be deposited only at an appropriately licensed tip. All removal or re-burial of contamination must be in accordance with the Control of Pollution Act (1974).

Buried structures are a hindrance to construction of new foundations and may prove costly to remove. Bridging across them, using piled foundations, may be an alternative. Piles, however, are expensive, and it may be necessary to check by exploratory drilling that pile locations are clear of obstructions. Piles will also require protection from aggressive ground.

The above recommendations presuppose that the site investigation has revealed conditions reasonably favourable to development. The conditions may be such that the developer may find it economic to build only on the relatively clean areas of the site, reserving the highly contaminated areas for car parks and similar non-sensitive purposes.

A3.1.12 Gasworks: further information

These notes dealing with gasworks are presented to illustrate the application of basic principles to particular types of industrially contaminated sites. The specific subject of gasworks has been thoroughly documented by Environmental Resources Ltd in considerably greater detail. Their report (HMSO, 1988) should be carefully studied by anyone concerned with the reclamation and redevelopment of old gasworks sites. Parker *et al.* (1984) give an interesting case history of a gasworks reclamation.

A3.2 Commercial sites

A3.2.1 General

This category covers light industry such as plating works, timber yards, warehouses, and scrapyards. Table A4 lists a number of typical commercial sites and associated hazards.

Initial assessment of all types of commercial site should be executed in broadly the same way, with an emphasis on early site visits, preferably before any demolition takes place. Unexpected sources of pollution can be present, which may be difficult to detect after demolition and clearance.

Scrapyards are considered below as representative of commercial sites. Much useful information on scrapyards is also given in ICRCL Note 42/80, 2nd edition (1983).

A3.2.2 Scrapyards: contamination levels

Contamination of scrapyard sites is likely to be high, with a wide range of possible toxic substances including organic compounds as well as metallic cations. Contamination levels

Table A4 Commercial works and associated hazards

Commercial works	Associated hazards
Scrapyards	See Section A3.2
Abbatoirs	Anthrax. Effluent may be contaminated
Small plating works	Chromium, cyanides, cadmium, nickel. (Generally necessary to isolate, remove and treat for disposal)
Warehouses	Many types of contamination possible. Historical survey should be thorough to reveal any special problems
Timber yards	Arsenic contamination from preservatives such as copper chrome arsenate, phenols, creosote, etc.
Coal yards	Combustion. Massive foundations
Railway land	As for coal yards plus water-treatment chemicals, oils, ash and metals
Farms and nurseries	Corrosive effluent from manure heaps, especially from intensive farming. High BOD and COD values. Ammonium salts. Copper in pig slurry
Printing works	Pigments can contain toxic metals, oils, organics
Commercial garages	Oils, paint pigments, phenols (from burning rubber tyres). Buried petrol tanks. Asbestos
Small manufacturing works	Almost any type of contamination could be present. Check historical records

Table A5 Maximum metal concentrations (mg/kg dry soil) for three scrapyards (after ICRCL, 1983)

Site	Zn	Cu	Ni	Cd	Pb	Hg	Se
Typical agricultural soil (normal max. values)	300	100	500	1	200	0.3	0.5
Scrapyard A	12 000	9 000	350	50	7 400	0.4	
Scrapyard B	7 000	10 000		1 500	20 000	1.0	
Scrapyard C	56 000	16 000	3 000	130	12 000		60

Note: Concentrations exhibit wide range.

can vary widely as illustrated by Table A5, which shows the maximum concentrations of metallic ions measured at three separate scrapyards, compared with the maximum corresponding values found in typical agricultural soils.

A3.2.3 Scrapyards: sources of contamination

Scrapyards may be contaminated by finely divided solids, or by liquids from spillages and leakages, or deliberate pouring away of waste, or through leaching of soluble contaminants by surface water. Scrapyards may be small in area, but the extreme heterogeneity of the contamination can result in toxic materials not being detected. Knowledge (if trustworthy) of the form of tipping, breaking or other activity involving scrap can give an indication of the type of contamination that may be present. An early visit to the site before clearance (usually by burning and by baling metallic residues) can be illuminating, and may prompt better disposal methods designed to reduce final contamination of the site or of the surrounding area by airborne pollution.

Typical activities leading to surface and sub-surface contamination of scrap yards include:

- Dismantling of motor vehicles and aircraft
- Storage and dismantling of plant and machinery
- Metal cleaning, sorting and recovery, particularly non-ferrous metals
- Recovery of electrical components from equipment and appliances
- Burning of cables to recover copper and of vehicle batteries to recover lead
- Crushing and fragmentation of bulky ferrous scrap to provide feed for blast furnaces
- Burning of rubber tyres and other inflammable materials to reduce bulk.

Cleaning operations result in dumping of contaminated sludges, release of fluids contained in internal combustion engines, transmission parts, accumulators, transformers and capacitors. These produce a range of contaminants including:

- Metals: arsenic, antimony, barium, cadmium, chromium, copper, iron, lead, manganese, mercury, nickel, tin and zinc
- Non-metals: mineral oils, cyanides, fluorides, phosphorus, sulphides, sulphates, PCBs, solvents (used to clean recovered parts).

Conditions in scrapyards may be aggravated by general untidiness and lack of control over the activities carried out on-site, including possibly fly-tipping and illicit disposal. Oils (commonly from car-breaking) may be particularly troublesome. These can seep into soil and penetrate gravels 2 to 3 m thick to form layers at impermeable boundaries, leaving the gravels apparently clean. These layers of oil may squeeze to the surface later or may attack plastics used for services.

A3.2.4 Scrapyards: characteristic hazards

Geotechnical problems are usually limited to the properties of the natural soil unless there has been infilling, which should be revealed during the site investigation.

Contamination, on the other hand, gives rise to some or all of the hazards discussed in Section 12.

A3.2.5 Scrapyards: site investigation

Trial pits are useful in revealing sources of contamination. Boreholes will be required for checks on contamination of groundwater at depth and for the installation of piezometers. Samples of surface and near-surface soil are especially relevant as well as those of any extant vegetation.

Scrapyard sites are usually small but contamination generally widespread. Therefore preliminary sampling should be on a square or rectangular grid of about 25 m, depending on the size of the site, but may be smaller for a secondary, more intensive, study of particular areas.

A3.2.6 Scrapyards: basic chemical analyses

The chemical analyses required will depend on the proposed end use of the site. The chemist should formulate the analytical programme and should be present to observe materials *in situ*, especially where there are uncertainties about the composition of substances and levels of risk to the investigation team. Mineral oils, coal tars, PAHs and PCBs should always be determined.

As a basic guide, the other chemical analyses likely to be required may be restricted to identifying some of the following 'listed' contaminants (Table 15) depending upon the planned end use:

Copper*	Cadmium
Nickel*	Lead
Zinc*	Sulphate
Chloride	Sulphide.
Arsenic	

The pH value should also be determined. Radioactivity is also a hazard on some scrapyards.

The 'plant-available' concentrations of copper, nickel and zinc (asterisked) are of importance if a sensitive end use is planned. The determinations of pH, chloride, sulphate and sulphide should be carried out on both the solid samples and on water extracted from them where appropriate. Samples of groundwater or other liquid encountered, as well as surface vegetation, if any, should be collected for analysis.

A3.2.7 Scrapyards: additional chemical analyses

Given the wide variety of activities and materials handled on scrapyards, additional contaminants beyond the 'listed' items may be present. Table A6 gives an extended list of contaminants that may be encountered on a scrapyard, and which may be important to some end uses. However, the advice of the chemist should always be sought before further analysis is contemplated beyond that proposed in Section A3.2.6 above.

Table A6 Additional contaminant hazards encountered on scrapyard sites

Metals		
	Antimony	Molybdenum
	Boron	Selenium
	Mercury	Vanadium
	Chromium (3- and 6-valent)	
Acids		
	Hydrochloric	Sulphuric
	Phosphoric	
Alkalis		
	Caustic soda	Ammoniacal liquors
Organic compounds		
	Cyclohexane	Cleaning/degreasing liquids
	Solvents/PCBs	Soluble compounds
	Toluene	
Various		
	Plastics	Rubber
	Glass	Metal scrap
	Coal wastes	Metal dusts

Note: These items are additional to the basic 'listed' hazards set out in Table 15. Not all will be important on a given scrapyard site, and for several of them ICRCL 'trigger values' are not yet available.
Site history may point to the likelihood of encountering a particular contaminant, but expert scientific advice will be needed (see also Section A3.2.6).

A3.2.8 Scrapyards: assessment and treatment

Analysis results should be plotted so as to map zones of contamination, using maximum values of concentrations, and an appraisal made of the hazards. Given the typically small size of scrapyards there may be little scope for modifying the proposed layout (i.e. planning option 3, Section 17.2).

Where the contamination is localised in small zones, its complete removal and replacement may be feasible, or it may be re-buried on-site. Alternatively, it may be diluted by mixing with clean site material and redispersed on the site. During this operation acidity may be corrected with lime.

The waste materials will probably be notifiable as Special Wastes under Section 17 of the Control of Pollution Act 1974 so that the consent of the waste-disposal authority will be required, either for their removal or their re-burial on site.

The depth of contamination will influence the choice of treatment. There is little point in exposing contamination at a depth when the surface is only lightly contaminated; a capping layer may be more appropriate.

Over-riding constraints, such as danger to water supplies or ground levels unsuitable for capping, may necessitate complete removal of all contamination. Precautions will probably be required against aggressive attack on building materials and services.

A3.3 Domestic sites

A3.3.1 Residential sites

Typically, in inner-city areas, these are densely packed late-Victorian dwellings demolished for new building works. Extremely local concentrations of contaminants may exist where small-scale industries were confined to individual houses, for which no historical records may now be found. For example, in the London Borough of Greenwich high levels of mercury were found in a back garden once occupied by a dental mechanic. Unless excessively fine-resolution sampling is to be carried out, there is little certainty of catching such small pockets of contamination. This can be embarrassing if sensitive end uses are contemplated.

Old street directories may indicate non-residential uses. Urban garden soils are also, in fact, typically contaminated with lead, zinc, mercury, etc. because of aerial deposition, dumping of coal ash and soot, burning of refuse on bonfires, flaking lead paint, and breakdown of galvanising.

Otherwise, in the absence of firm information on past history, the most effective approach is probably a preliminary surface and limited sub-surface sampling from trial pits, at about 25 m spacing, backed up by a secondary investigation based on the findings of the first. Basements are commonly present and need careful checking, particularly if the site has been unfenced for some time in a derelict state. Fly-tipping may have taken place with basements used as dumps for toxic materials or builder's rubble containing asbestos. Demolition rubble containing gypsum from plaster can produce elevated sulphate levels in soil if leaching by rainwater has been prolonged.

The basements themselves are a hindrance to new foundation construction, and an investigation of the location, depth and extent of these structures should be made. The extent of such investigation would depend on the proposed end use of the site. In the 1960s much redevelopment involved high-rise blocks built on piles driven through old basements, with pile caps at basement level. Settlement was thus not a problem, and the basements were either excavated completely or backfilled to provide a platform for contractors' plant. If, on the other hand, low-rise development is proposed without piling, the settlement characteristics of a site backfilled in patchwork pattern will be important.

A3.3.2 Hospital sites

Potential problems requiring investigation on old hospital sites include:

- Presence of old foundations and deep basements
- Drying-out of clays beneath furnace floors, with possibility of future swelling
- Contamination by pathogens or pharmacological waste
- Widespread contamination resulting from dismantling of asbestos lagging on pipework.

Identification of boiler houses, incinerators and waste dumps is clearly necessary. This is probably easier for hospital sites than for industrial sites. Old records should be studied in case there has been an interim redevelopment. If asbestos waste is widespread, a painstaking examination of the surface on a close grid will be necessary, sufficient to decide whether capping is acceptable or whether complete removal is required.

A3.4 Opencast fill sites

A3.4.1 General

Opencast mining has been practised in the UK mainly for the extraction of shallow coal deposits. Other minerals similarly extracted include ironstone (extensive at Corby and elsewhere in Northamptonshire), china clay (extensive in Cornwall), fire clay and cellestine. Opencast backfill is often relatively free of biodegradable material and contamination.

Opencast coal workings generally cover many hectares and constitute the most extensive sites likely to be developed. The present trend is to yet deeper and larger pits. It is estimated that, for coal mining alone, 200 million m^3 of overburden are stripped every year.

Complications derive from existing underground workings, weathering of shales, the carbonaceous content of the backfill and large-scale pumping in the vicinity. Opencast mines are typically at least 30 m deep, with the mined seams generally inclined at low angles of dip (15% or less). The strata to be mined may be underlain by similar seams, already worked by traditional underground methods.

A3.4.2 Opencast mining operations

A typical opencast operation is to excavate in cuts running parallel to the strike. During the period that the mine is operating, the first backfilled spoil (from the second cut) has the longest time to settle under its own weight and may be compacted by site traffic. Such traffic is likely to be localised along haul roads, etc. and variable in intensity.

Other working methods are possible. Where there is risk of sliding on dip planes, the cuts may be parallel to the dip of the seam, rather than the strike. For deeply inclined beds, on the other hand, the 'open pit' method is used to depths in excess of 100 m, whereby an excavation is made to expose the seam at the base of the pit and deepened as the seam is worked. Differing patterns of excavation, stockpiling and backfill result from the various methods.

The mining method and the type of plant strongly influence the settlement to be expected. The plant usually consists of dragline and scrapers, or dragline with face-shovel and dump trucks. In the reinstatement of opencast sites the spoil heaps are regraded and the topsoil replaced.

Prior to about 1965, uncontrolled backfilling by dragline, or end tipping from wagons, was practised; and the resulting fills were usually restored to agriculture. They have since been settling under their self-weight only, and their upper layers may still require some densification before being used for building works.

More recently, on sites earmarked for future building development, the upper 15 m or so have been compacted, either by spreading and rolling in thin layers of 150–300 mm or, more cheaply, by laying with scrapers. Knipe (1979) reports that the settlement behaviour of scraper-laid fills compares favourably with that of spread and rolled fills, provided that the reinstated site is not required for development within three years.

A3.4.3 Properties of opencast backfills

In the UK, the Coal Measures overburden generally comprises sandstones and grits, with mudstone, shales, siltstones and seat earths. Excavation reduces the overburden to particle sizes ranging from below 200 microns to large boulder size. The excavated rocks are susceptible to rapid weathering as result of stress relief combined with exposure to air and water. Superficial argillaceous material tends to mix with the stripped bedrock, causing erratic alteration of the settlement characteristics and permeability of the backfill. Bulk densities are generally in excess of 2.0 mg/m^3, with varying moisture contents up to full saturation.

A feature of opencast coal backfill is the residual bulking after restoration (about 5%). This is an advantage in landscaping and in establishing surface drainage away from the fill area.

Opencast backfills are sensitive to the water table level, and can suffer 'collapse' settlement if the groundwater rises. The collapse is most severe following the first rise in the water level; subsequent rises have comparatively minor effect (Leigh and Rainbow, 1979).

A3.4.4 Opencast sites: historical and other investigations

The desk study of a backfilled opencast coal site should investigate the following items for the site and the surrounding area:

- Old workings
- Current workings
- Silt and sludge lagoons and coal washing plant
- Drainage and pumping arrangements in underground workings in the vicinity of the site (including future pumping programmes)
- Future underground workings proposed for driving near to or under the site
- Associated coking and chemical plants and location of any residues.

For these items British Coal should be consulted.

Surface settlement of the backfill is the greatest problem. Ideally, monitoring of ground and water table levels should be started as soon as possible after reinstatement and continued for at least 12 months. Records (if any) of such monitoring are an important asset to the desk study.

Historic records of the excursions of the water table are important in determining the susceptibility of the fill to future collapse settlement. In the absence of such records, it is at least essential to be sure that the water table is standing at its highest probable future level. This requires careful judgement by qualified experts.

A3.4.5 Opencast sites: assessment

The information required on an opencast backfill site includes:

- Extent of site and boundaries of disturbed ground
- Depths, and changes in level of bedrock liable to cause differential settlement
- State of compaction of fill material
- State of surface settlement
- Fluctuations in water table
- Possible chemical attack (presence of sulphides and/of low pH)
- Potential combustibility.

The first two items should be initially deducible from a comparison of the regraded ground levels and bedrock levels shown on the opencast excavation records.

A3.4.6 Opencast sites: site investigation

Boreholes through the full thickness of the fill will be required for groundwater observations and *in-situ* permeability measurements in areas to be developed. These tests can indicate fill variability and settlement characteristics. The required spacing of boreholes is related to the variability of the fill and type of development proposed. A broad coverage on a 100–200 m grid should normally be adequate with closer spacing near proposed foundations.

For guidance on the economic location of boreholes, the site investigation may include a prior geophysical survey, using resistivity methods for identification of the excavated volume. Anomalies in resistivity may indicate major variations in fill properties.

Boreholes would generally be advanced by cable percussion boring. Air-flush drilling is a possible alternative for proving rock head, high walls, mine shafts, adits and workings beneath opencast workings, as well as for *in-situ* testing. Samples should be used for identification of the composition and condition (state of weathering) of the components of the fill, its particle size distribution, bulk density (if possible), moisture content, pH and sulphate content (also on water samples if available). Proctor compaction tests can be useful for comparison if representative bulk samples are used.

The borehole information should be supplemented by Standard Penetration Tests (SPT). In homogeneous, fine-grained backfill with SPT N values less than 50, pressuremeter tests provide additional information on soil-deformation properties, but their economic justification is questionable.

The condition of material at shallow depth (to about 5 m) is best inspected from trial pits, from which bulk samples are more likely to be representative than those from boreholes. This coincides approximately with the effective depth of ground improvement by dynamic compaction.

If the land is required for building, settlement investigation will be necessary. For assessment of the creep settlement potential of the backfill under self-load and its susceptibility to collapse settlement, ground level and water table monitoring should be put in hand at the earliest possible moment and continued for an extended period. Backfills placed fairly recently have often been spread in layers. For such fills a ground-level monitoring a period of 12 months would probably be sufficient. Older backfills were not usually placed with such care. Even with allowance for the longer period since deposition, a monitoring period of 18 months is advisable.

Water table levels should be recorded in boreholes for an extended period, possibly with automatic recorders.

A3.4.7 Opencast sites: main problems

The main problems when building on backfilled and regraded opencast sites are as follows:

- Large total settlement
- Large differential settlement, especially near edges of pits across the high wall or in the vicinity of haul roads. The compressibility of the fill varies as a result of the sequence of stripping and dumping operations
- Heave where large stockpiles have been regraded
- Sudden increased settlement if inundation occurs subsequent to building
- Downdrag on piles
- Chemical attack. Depending on the nature of the opencast site with coal or metalliferous ores, sulphide may be present and the pH low.

A3.4.8 Opencast sites: treatment methods

Monitoring of settlement may show that the fill does not need further treatment. If the degree of compaction of backfill is deemed inadequate, dynamic compaction or vibro methods may be used to improve the top 5 or 6 m. Vibrated stone columns may be used to any desired depth, provided the fill is free of obstructions and significant amounts of domestic refuse. Whether or not the fill so treated will provide a satisfactory foundation depends on age, depth, and physical characteristics of the fill underneath the treated layer.

Structures of large extent are better not located on deep fill. If they are an indispensable part of a development plan, piling may be the only solution.

For economy, dynamic compaction, vibratory treatments or piling should be limited to the areas of the buildings requiring support, although low-energy dynamic compaction may provide economic treatment to adjacent areas of infrastructure works. For light buildings of small extent on stiff raft foundations, excavation and recompaction of the fill down to a nominal 3 or 4 m may prove more economic than dynamic compaction.

Differential settlement may be serious between buildings on piles and services in the ground. Large-diameter sewers may need support on piles, especially if they continue on to undisturbed ground. Smaller-diameter pipes should be flexible or have joints permitting deflection, and should be supported from ground beams within the confines of a building. Drains resting directly in the fill should be laid to a slope of not less than 3% to avoid risk of backfall. Cables may have sufficient flexibility to accommodate settlement, provided sufficient spare length is available to allow for ground strains.

No reports have been received of combustion occurring in opencast coal-site backfill. It is generally believed that any residual carbonaceous matter is too diluted by inert material during the operation of backfilling to constitute a hazard.

Appendix 4 Systematic gas investigation

Carpenter *et al.* (1985) Pecksen (1985) and Barry (1987) have dealt with the criteria, procedures and interpretation of results for the systematic investigation of gas concentrations and flows in the ground. This appendix is based on their publications. For details of available instrumentation and methods of measurement reference should be made to Crowhurst (1987).

A4.1 Scope of investigation

If the site history suggests the presence of putrescible waste in a fill a gas investigation will normally be necessary, irrespective of the age, type or quantity of the fill, before any end use involving building works can be contemplated (Figure A2). Given the natural tendency of gas to migrate laterally, the investigation should extend beyond the periphery of the fill area. Reclaimed estuarial land, dockland, etc. should also be investigated for gas potential wherever silt, peat or organic deposits are suspected.

Figure A2 *Refuse dated at 1965 and still capable of producing methane (recent photograph by courtesy of M. A. Smith)*

A gas investigation is usually conducted in two stages: a preliminary sub-surface 'spiking' survey to help identify the locations for the deeper second-stage systematic investigation. The spiking survey can provide valuable indications of transitory sub-surface concentrations and gas-migration tendencies, but is not a substitute for the deep investigation.

The aim of an investigation is twofold:

1. To produce a contour map of gas concentration in the ground. Given the mobility of gas, this will be valid for the date of sampling only, and will require repetition, as outlined below
2. To determine the emission rate of gas at determined points. From this there are empirical methods for the approximate computation of generation rates of gas in the ground and values for the design of remedial measures (see Pecksen, 1985, and Section 17.8.4).

A4.2 Preliminary investigation

This involves a spiking survey to determine the sub-surface (500-mm depth) concentration of gas at a large number of regularly spaced points. A portable gas detector probe is inserted into holes formed by driving a 12-mm diameter metal bar 500 mm into the ground (Figure A3) on an accurately surveyed 10–25 m grid. The gas detector should register with accuracy concentrations as low as 10 ppm methane and up to 100% methane. Measurements of carbon dioxide should be made as well.

Figure A3 *Sampling landfill gas (photograph by courtesy of GLC)*

From the results a preliminary concentration contour map can be made. However, the interpretation of spiking surveys is not straightforward. If any methane or higher than normal CO_2 levels are detected, further investigation and specialist advice are needed.

A4.3 Deeper investigation

The aim of the deeper investigation is to produce a dynamic profile of gas distribution throughout the whole depth of the site. Due to the variability of the gas, tests should extend over at least two weeks, and then be repeated at monthly intervals for six months.

On-site measurements should always be made for methane, carbon dioxide and oxygen; and samples should be taken for laboratory analysis to check the site measurements.

Boreholes or probes should be located on a grid at the most strategic points of interaction between the proposed end use and the suspected gas concentrations as indicated by the spiking tests. Spacings may be 25×25 m or, more economically 50×50 m (spacings of more than 75 m are usually of little value), with more closely spaced holes in areas proposed for concentrated development.

Boreholes should be sunk by cable percussion techniques to a depth of at least 6 m, regardless of the water table, with gas sampling from the bottom every metre during driving. Upon completion of the bore, a perforated plastic tube should be inserted into the casing, with granular fill in the annular space, and the steel casing should then be removed. Recommended borehole details are as shown in Figure A4. Similar monitoring installations can be placed in backfilled trial pits.

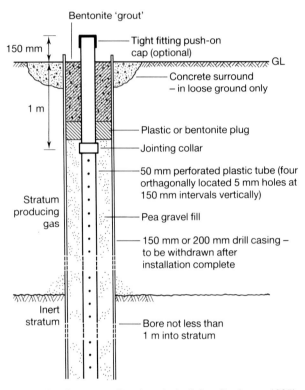

Figure A4 *Gas-sampling borehole (after Pecksen, 1985)*

Gas emission rates (measured in m/s) from the borehole should be measured upon completion of the bore, and repeated after 24 hours and again after two weeks. Gas samples for chemical analysis should be taken at the same time by the procedure set out below.

It is of great importance that gas emission rates be measured with an instrument accurate at low rates, for instance, a sufficiently sensitive hot-wire anemometer. This may require the use of more than one grade of instrument where large flows are expected in certain parts of the site. Accuracy at high rates may not be critical if the rates are large enough to rule out the site for development.

A4.4 Gas-sampling methods

The accurate determination of the emission rate from the borehole, especially in the case of very low emission flows, requires first purging the borehole with air, or preferably nitrogen, and observing the subsequent rate of build-up of gas concentration (Figure A5). To that end, samples are to be taken at regular intervals during the recharging of the borehole for analysis by gas chromatography.

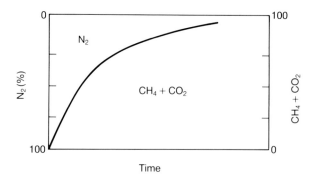

Figure A5 *Recharge of landfill gas borehole after purging with nitrogen*

Recharge of the borehole is an equilibration process between the borehole atmosphere and the ground atmosphere, and decays exponentially. Intervals between sampling should therefore be short, and the overall sampling time a matter of minutes rather than hours.

A4.5 Interpretation of results

The final analysis and interpretation of results is complex, and all data must be viewed in conjunction with ground conditions (hydrology, geology, permeability, rise/fall of water table), atmospheric temperature and pressure at the time of sampling, as well as the proposed end use.

Expert advice is therefore needed for the interpretation of gas investigations, involving such factors as:

Concentration: equilibrium concentration in the ground
Emission rate: rate of gas emission from borehole (or fissure)
Design rate: rate of accumulation of gas in a given design space
Generation rate: rate of gas production from unit mass or volume of fill

A4.6 Long-term gas monitoring

Probes in sampling tubes, sealed into boreholes or trial pits at the desired depths or driven into voids, may be connected to an on-site laboratory for long-term monitoring, preferably with automatic recording. Monitoring should continue for an indefinite period, together with monitoring of the corresponding ground water levels. (Gas production has been found to be maximum at a moisture content of around 40–60% (Section 12.11.3) in putrescible landfill material.)

A4.7 Estimation of future gassing potential

Chemical analyses of available carbon in the biodegradable content of the fill are useful in assessing the approximate potential for future gas generation (Pecksen, 1985). This is important when consistently low methane values have been obtained over an extended period (Section 17.8.3).

Index

acceptance criteria 29, 102
acid attack 44, 49, 63, 124, 172
adjacent sites
  gas monitoring 32, 126, 174–5, 192
  landfill gas 125
  leachate migration 65, 173
  protection from gas 128, 150
  site reconnaissance 27
aerial photography 75, 76, 86, 175
aerobic decay 53, 66
age of fills 64, 67, 74, 106–7
aggressive ground
  effect on materials 63, 93, 121–2, 216
  in fills 38
  pyritic shales 45–7, 95, 101, 123–4
  remedial actions 147, 153–4, 169–70
agricultural land
  formerly derelict 16, 37, 114, 126, 218
  hazards to 57
air-barriers 142, 143
airborne contamination 68, 117, 193
  site reconnaissance 78–9, 214
  vegetation 55, 101, 173, 217
air-flush drilling 83, 151, 219
alarm sensors 166, 175, 188, 189, 191
ammonia 56, 60, 66, 210
ammoniacal liquors 56, 209
anaerobic decay 57, 64, 66
anemometers 223
animals
  hazards to 49, 53, 54–5
anthrax 185
aquifers
  contamination migration 16, 173, 174, 192, 212
  leachates 63–4
  protection of 68, 139, 144
arsenic 52, 56, 117, 215
  in food 55, 212
asbestos 52, 53, 211, 217
  as contaminant 68, 102
  protection from 187, 188, 189
  trigger values 117, 185–6
ash content of refuse 38, 39, 74
Assisted Area 205

bacteria 149
barometric pressure
  effects of 63, 100, 184
barrier walls 32, 65
  gases 127, 128, 142, 150
  groundwater 129, 146, 148, 155
  see also cover layers; fire fences
basements 50, 136, 168, 217
bearing capacity 142
  compacted fills 135, 136
  fills 38, 40, 44, 110
Beckton Alps 140, 170
bentonite seal 68, 152, 153
benzene 54, 60
bio-barriers 139, 142, 143
biodegradable fills 43, 66, 74, 187
  settlement 43, 106, 171
biological hazards 185
blastfurnace slag
  engineering hazards 44–5, 95, 101, 123
  health hazards 56, 57
blue billy 210, 211
boreholes
  control water table 129

gas sampling 100, 127, 223–4
gas ventilation 151
groundwater sampling 94, 98, 99–100, 174
site investigation 83–5, 90, 211, 219
boron 57, 117, 158
botanist 22, 73, 181
bottom sealing 155
break layers
  pfa 157, 158
  soil suction 139, 142, 143, 144–5, 154
brick buildings 160, 161, 164
Building Act 70
Building Regulations 69, 70, 145, 166, 207
building waste 44, 49, 95, 123–4
buried structures
  hazards 17, 49–50, 51, 145, 211
  piling hazard 37, 50, 71, 160, 213
  site investigation 78, 82, 85, 89, 211

cadmium
  in plants 55, 116, 117, 212
  scrapyards 212, 213, 214, 215
  trigger values 117, 213
calcite 47, 95
capillary migration 63, 65
capping layer 150–1, 216
carbon dioxide
  characteristics 54
  generation of 53, 61–3
  phytotoxicity 57
  safety 166, 185
  site appraisal 124–6
  site investigation 222–3
carbon monoxide 53, 54, 60, 185
carcinogens 53, 54, 57, 184, 186
car parks 211, 213
casings 83, 84, 85, 98, 134
cathodic protection 170, 171
cement 45, 57, 158
cement-bentonite plug 83, 84, 98, 192
cement grouting 147–8, 153
chemical analyses
  coverage 91–3
  gasworks 211–12
  monitoring 173–4
  sampling scheme 93, 94
  scrapyards 215–16
  specification 91
chemical barrier 142, 143
chemical hazards
  protection 184–5
chemical injection 148
chemist
  during construction 181
  role of 22, 73
  site appraisal 73, 102, 118–19
  site investigation 87, 90–1, 100, 211–12, 215–16
children
  sensitivity of 52, 53, 54–5, 56, 93
chimneys 95, 186
chlorides 56, 63, 211, 215
chromium 117, 158
clay pits 37
coal residues 23, 120
coal tar 53, 56–7, 209
cohesive soils
  consolidation 40, 44, 71, 132, 165
  embankments on 165

cohesive soils (*contd*)
  ground treatment 133–4, 137
  soil suction 154
coking plant 68
collapse settlement
  constraint to project 67
  examples of 42–3, 113, 161
  fill appraisal 108, 112–13, 114
  opencast mines 42–3, 161, 218–20
colliery spoils
  acid run-off 45–6, 64
  expansion of 101
  oxidation 45–6
  self-heating 58, 121, 153
combustion hazard
  adding water 60, 152, 186
  categories of 120, 152
  design factors 165
  development abandoned 38, 93, 120, 151, 186
  development constraint 23, 67, 69, 70
  excavation 28, 152, 153
  gases generated 53
  monitoring 175
  potential materials 23, 58–60, 67, 119
  safety 28, 151, 152, 183, 186
  sampling 90, 101
  self-heating 58–9, 120, 121
  settlement 38, 43, 58
  site appraisal 119–21
  site investigation 90, 92, 93, 94–5, 101
  treatment 151–3, 165
commercial sites 213–14
compaction 106, 107–8, 121, 152
concrete and sulphates
  building wastes and slags 44, 45
  design measures 164, 169–70
  protection 153–4, 160
  site appraisal 122
consolidation 37, 44, 110
constraints on development 23–5, 36, 67–72, T2
  collapse settlement 67
  combustion 23, 67, 69, 70
  cost factors 71, 72
  gas hazard 16, 23, 76, 102, 125
  housing 22, 23
  metals 23
  radioactivity 68, 69
construction
  delays 43, 113
  hazards during 71
contact, skin 49, 158, 184
contaminated sites
  aggressive attack 63
  chemical injection 148
  combustion and fire 58–61
  concentrations 50, 52
  containment 68, 192
  cover layers 138–45
  corrosion 63
  disposal 147
  disturbance of 146
  downward migration 65, 174
  encapsulation 154–5
  excavation and disposal 32, 145–7, 153–4, 213
  exposure time 52
  gas emission 61–3
  groundwater 32, 55–6, 63–4
  hazards 48–9, 51, 208–9
  health hazards 50, 52–7
  in gasworks 210–12
  in-situ treatment 147–8
  lateral migration 65, 173–4
  leachates 63–4
  migration 18, 27, 103

monitoring 173–4
  neutralising 147, 148, 149
  off-site migration 68
  phytotoxic 57
  refuse fill 66
  sampling 87–90, 173–4, 211
  settlement 17, 218, 219
  soil processing 148–9
  soil suction 57, 65, 139, 158, 173, 210
  sources 21–2, 48
  upward migration 65, 90, 143, 153
  use of bacteria 149
contractor, analytical 90–1
contractor
  constraints on 36, 68, 179, 180
  experienced 180
contractor responsibilities
  contractual 177, 179–80
  safety 24, 71, 179, 182
  third parties 71, 179, 182
contracts
  content 36, 178–9
  form 178
  need for flexibility 177
  safety factors 179–80, 181
  specification 36, 179–80
  tendering 36, 179, 180–1
  types of 36, 177–8
  unforeseen hazards 36, 177, 178, 179
Control of Pollution Act
  disposal of contaminants 147, 192, 213, 216
  fill sites 37, 127, 150
  fly-tipping 48
  in Scotland 207
controlled combustion 66
copper 57, 117, 212, 214, 215
corrosion 63, 92, 93, 121–2
  design against 122, 169–70
  protection of services 171–2
corrosion scientist 22, 122, 169, 181
cost factors
  derelict land 24–5, 71
  expertise 22
  ground treatment 132, 133, 136, 137
  project constraints 71, 72
  remedial works 129, 145, 147, 151, 152
  safety factors 17–18, 71
  site investigation 79, 80, 81, 205
cover layers 65, 213
  design factors 142–5
  development 139–40
  fire hazards 120, 152
  functions 32, 141–2
  gas hazards 127–8, 129, 150–1
  layered sandwich 141
  leachate hazards 155
  materials 142
  multi-purpose 142, 144–5, 153
  water supply in 70
creep settlement 40–2, 106, 137, 219
  prediction 109–10, 112
crusts 37, 42, 44, 158, 187
cut-off barriers 127–8, 150, 151, 153
cyanides 52, 53, 56, 57, 117
  trigger values 117, 118
  see also hydrogen cyanide
cyanogen compounds 55, 210

demolition 22, 64, 68, 82, 186
Deposit of Poisonous Wastes Act 74
derelict land
  definition 15–16
  map of England 15
Derelict Land Grant 205–6

Derelict Land Survey 15
dermatitis 56–7, 158
design factors 30, 34
    against settlement 104, 159–65
    aggressive attack 122, 169–70
    buildings and gas 165–9
    combustion hazard 165
    corrosion 122, 169–70
    cover layers 142–5
    piling 113, 159, 160–1
    protection of services 170–2
    slags and expansive materials 172
see also foundations
desk study 27, 74–6, 210–11, 218
developer responsibilities
    flexibility 27, 103
    funding 103, 177
    safety 17–18, 69, 71
    third parties 17–18, 32, 71
    unforeseen hazards 177, 178
    waste disposal 93, 147
development
    delays to 43, 113
    inadvisable 67–8, 125–6
development abandoned
    acceptance criteria 29, 102
    action required 32
    fire hazard 38, 93, 120, 151, 186
    gas hazard 38, 80
diaphragms 146
differential settlement 31, 34, 42, 130, 220
    factors affecting 114
    foundation design 159–64
    monitoring 176
    predicting 104, 108
    reduction of 131
dioxins 53, 68, 102, 117, 174
disposal of contaminants 147, 192, 213, 216
distortion of buildings 105, 106, 159, 160
docks 74, 124, 127, 164, 221
domestic refuse
    biodegradation 66, 156
    constituents 38, 39
    gas hazard 61, 67, 124
    moisture content 62
    settlement 43, 67, 114
downdrag 38, 113, 160, 220
drains
    gradient 220
    leachate protection 63, 129, 154, 156
    protection of 146
    road 164
    water table lowering 63, 129, 153, 155
drilling 83, 151, 219
dust
    health hazard 52, 53, 158
    off-site 146
    site workers hazard 49, 145, 184
dynamic compaction 115, 131, 136–8, 219–20

earthworms 57, 78, 139, 143
elderly and infirm 53
electricity companies 32, 69, 170
electolytic action 49, 93, 154
embankments 165
emergencies 32, 69, 79, 191
end use
    changed 29–30, 103
    chemical analyses 91–3
    factors affecting 19, 21, 23, 91
    procedure 19–20
environment, impact on 128–9, 146
environmental health officer 16, 71, 158

Environmental Protection Act 69
environmental responsibilities 17, 23–4, 32, 71, 102
environmental scientist 22, 73, 77, 181
excavation
    contaminated sites 32, 145–7, 153–4, 192, 213
    fill sites 31, 113, 115, 132
    safety aspects 28, 145, 186, 188, 189, 191
expansive fills 44–5, 46, 47, 95
    sampling 95, 101
    treatment 123–4, 138, 164, 169, 172
explosions 60–1, 63, 69, 93, 186
explosive limits 60
exposure time 52
extensometers 176

fill sites 16, 21, 37–8, 74–5
    age 64, 67, 74, 106–7
    biodegradable 43, 66, 74, 187
    classification 105, 113–14
    compressibility 107–8, 114
    construction on 104, 115
    contamination 38, 44–7
    depth 107
    depth of treatment 131, 135, 137
    dynamic compaction 136–8
    embankments on 165
    excavation 31, 113, 115, 132
    expansive 44–5, 46, 47, 95
    grading 107
    grouting 148
    imported 156–8
    methods of densifying 130–1
    preloading 131–2
    protection of services 171
    remedial work 31–2, 34, 130–8
    roads 164–5
    settlement 43, 105–13, 171
        monitoring 75, 77, 81, 165, 175–6
    unsaturated 40–3, 137
    vibratory compaction 133–6
filter layer 142, 143
finance of development 24–5, 72, 103, 205–6
fire extinguishers 189
fire-fences 120, 132, 152, 165
fire hazards
    flammable materials 60, 78, 183, 192
see also combustion hazards;
        underground fires
Fire Research Station 119, 121
fire service 32, 69, 186
first aid equipment 189
fish, hazards to 53, 55
flammable substances 60, 78, 183, 192
flash point 60
floor slabs 138, 163–4, 167–8, 169, 172
fly-tipping 48, 70
foundations 23, 34, 104, 123–4
    choice 115, 159–64
    protection of 153–4, 169–70
    type of contract 177, 178
see also piling; rafts; strip footings
framed structures 159, 161, 164, 172
fumes 52, 53, 78, 145, 146

gardens
    gas hazards 125
    health hazards 55, 57, 92–3, 117, 118
    planning constraint 22, 23, 213
    topsoil 57, 157
gas hazard
    barriers 142, 143
    characteristics 54
    classification of site 125–6

gas hazard (*contd*)
  control of 127–8, 149–51
  design of buildings 128, 165–9
  detection 26, 78, 100, 188, 222
  disturbance 62, 137
  drainage 142, 143, 150–1, 166, 168
  extraction 126, 127–8, 150–1, 166
  flaring 150, 151
  generation rate 126–7
  inhalation 53–4
  measuring emission 127
  migration 33, 62–3, 127–8, 221
  monitoring 63, 81, 125–6, 191–2, 223–4
    after construction 166, 174–5
  occupational exposure limit 185
  planning constraint 16, 23, 102, 125
  production process 53–4, 61–3, 66
  safety 174, 184–5, 188
  sampling 87, 89, 100, 127, 174–5, 222–4
  site appraisal 102, 124–8
  site investigation 96, 221–4
  spiking survey 78, 89, 100, 222–3
  ventilation 125, 127–8, 150–1, 166–8, 175
gas suppliers 32, 69
gasworks sites
  buried obstructions 49–50, 78, 145
  chemical analyses 211–12
  combustion hazard 58, 119
  groundwater 55, 212
  health hazard 55, 56
  initial assessment 210–11
  production process 209–10
  remedial treatment 149, 213
  site appraisal 212–13
  site investigation 89, 211
  water supply hazard 55, 212
Gateshead gasworks 50, 140, 145
geochemical barriers 53, 55
geomembranes 143, 164, 166, 171
geophysical survey 85–6
geotechnical engineer 22, 73, 77, 105, 181
geotechnical tests 86–7
grants 16, 24, 72, 205–6
gravel pits 16, 37
groundwater
  barriers 129, 146, 148, 155
  contamination 32, 55–6, 63–4
  gasworks site 55, 212
  migration 33, 148, 153, 173–4
  monitoring 81, 84–5, 94, 174, 212, 219
  sampling 94, 98, 99–100, 174, 212
  site investigation 93–4, 98–100, 212
grouting 147–8, 151, 153, 155, 160
gypsum
  expansion 47, 95, 123
  gas produced 53
  sulphate attack 44, 217

hazards, summary of 21, 92–3, 208–9, 214, T2
Health and Safety at Work Act 69
Health and Safety Executive 182, 186, 192, 193
health hazards 50, 67, 93, 102, 187
  availability of contaminants 52
  contact 56–7, 158, 184, 186
  exposure time 52
  ingestion 49, 54–6
  inhalation 49, 52–4, 158, 184, 185
  trigger levels 102, 116–19, 212, 216
heave 45, 160, 172, 220
  after ground treatment 132, 137–8
heavy metals 52, 62, 78, 148, 184
  planning constraint 23
  soil suction 65

herbicides 56
Hillingdon, Willow Tree Lane 104, 140
Horsley opencast mine 42, 112
hospital sites 217
housing
  gas hazard 125, 166–8
  foundation choice 159–61, 172
  ground treatment 135, 138
  on pfa 157
  planning constraint 22, 23
housing, disused 74, 216–17
hydrocarbons 52, 56, 118, 172
  polynuclear aromatic 56, 118, 211, 215
hydrogen
  from buried metals 184
  in landfill gas 61
  in town gas 209
  properties of 54, 60, 185
hydrogen cyanide 53, 54, 185
hydrogen sulphide 68, 100, 210
  generation 44, 53, 61, 66, 184
  hazard 57, 185
  properties 54, 60, 185
hydrogeologist 22, 94, 113, 129, 181
hydrogeology survey 83–5, 153, 212

ignition point 58
ignition temperature 60, 61
imported fill 132, 156–8, 165
industrial sites
see contaminated sites
industrial specialist 22
industrial waste, treated 157
infiltration barriers 139, 142, 143
infra-red analysis 78, 86, 95, 100, 175
ingestion, hazards from 49, 54–6
inhalation, hazards from 49, 52–4, 158, 184, 185
initial assessment 19, 26, 73, 80
  desk study 27, 74–6, 210–11, 218
  gasworks 210–11
  site reconnaissance 27, 77–9, 182, 214
insulation layer 142, 143
iron-making slags 44–5, 50, 95, 101

jarosite 46–7, 95
jet grouting 148, 155

katharometer 78, 100
Keene's cement 44

laboratory, location of 100
lagoons
  from aerial survey 75
  pfa 37, 158
  safety 158, 187, 192
  settlement in 44, 67, 108, 114
landfill gas
  control of movement 126–8, 149–51
  decay process 53, 61–2, 66
  monitoring 174, 224
  phytotoxicity 57
  planning constraint 23, 76, 125
  properties 54, 61
  site appraisal 78, 124–8
  site investigation 90, 96, 221–4
landscaping 115–16, 132, 143, 149
layout, change of 23, 29, 103
leachates 53, 214
  environment threat 23, 128–9
  landfill sites 61, 63–4, 66
  migration 64, 70, 76
  monitoring 66, 84, 85, 94, 174
  remedial works 129, 147, 154–6

site investigation 84, 85, 87, 93–4
waste disposal 129, 156
water supply hazard 23, 38, 63, 70, 128–9
lead 53, 117, 212, 215
dust on vegetables 55, 173, 217
legal action 25, 71
levelling, precise 176
lime
free 45, 124
neutraliser 148, 149, 216
skin irritant 57
lining walls 150
liquefaction 44, 114
liquified petroleum gas (LPG) 60, 185
load tests 35, 110–12, 135, 136
local authorities
as land developers 205–6
gas hazards 150, 151, 192
in Scotland 207
responsibilities 16, 32, 69–70, 102
waste disposal 68, 129
loss on ignition 121
louvred walls 167–8
lower explosive limit 125, 166

magnesium oxide 45
maximum explosion pressure 60–1
medical services 69, 102
mercury 55, 117, 214, 217
metals
contaminants 49, 52, 56, 57, 78
plant essentials 49, 57
see also heavy metals
methane
building hazard 125, 126
landfill gas constituent 53, 54, 57, 61–3
lower explosive limit 125, 166
properties 54, 60, 125
site appraisal 124–7
site investigation 89, 221–4
treatment 125, 127–8, 166
microbiologist 73, 102
microbiological treatment 149
migration of contaminants
combating 153, 192
downward 65, 174
gas 33, 62–3, 127–8, 221
groundwater 64–5, 70, 153, 192, 210
lateral 65, 173–4
leachates 64, 70, 76
methods 64–5, 68, 210
off-site 68, 209
site investigation 88, 90, 100
soil suction 57, 65, 139, 158, 173, 210
tars 65, 67, 139–40, 141
upward 65, 90, 143, 153
monitoring
boreholes 63, 81, 84–5, 174, 223
contaminant migration 32, 173–4, 192
early 79, 81, 83, 84, 173
gas emission 63, 81, 125–6, 191–2, 223–4
after construction 166, 174–5
spiking survey 78, 89, 100, 222–3
ground temperature 81, 95, 153, 175, 186
groundwater control 155–6
groundwater level 81, 84–5, 94, 174, 212, 219
groundwater quality 81, 174
leachates 66, 84, 85, 94, 174
radioactivity 185
reasons for 20, 35, 173
settlement 75, 77, 81, 165, 175–6
predictions from 108–9, 110–12
water supplies 174, 212

multi-disciplines, need for 19, 22, 73, 179
multi-purpose layers 142, 144–5, 153
multi-spectral imagery 86

narcotics 54, 185
National Radiological Protection Board 69, 79, 101, 158
safety aspects 185, 192
National Rivers Authority 69, 70, 76
nickel 57, 117, 212, 214, 215
nitrates 66
noise pollution 78
nuisance 23, 32, 78, 151, 180

obstructions
see buried structures; piling
occupants of site
health hazards 49, 52, 53
maintenance 182, 183
safety responsibilities 17, 24, 69, 71
oils
combustion hazard 119
health hazard 49, 57, 187
squeezing out 65, 67, 210, 215
opencast mines 37, 85, 108, 120, 217–20
backfilling method 16, 107, 114, 218
oxidation of shales 45–7

PAHs (polynuclear aromatic hydrocarbons) 56, 118, 211, 215
paper refuse 38, 39, 62, 66, 221
PCBs (polchlorinated biphenyls) 174, 215
health hazard 52, 56, 57, 184
penetromter tests 86–7
peat 124, 221
perched water tables
in fills 38, 84, 85
water quality 55, 64, 144
periclase 45
permeability of ground 63, 100
pesticides 56, 148
phenols
gasworks sites 209, 211
health hazard 52, 56–7, 117, 118
in drinking water 55, 172
phytotoxicity 57, 78, 116, 212
piezometers 83, 85, 212
piling 32, 147, 167, 179
aggressive ground 123, 169, 172, 213
design factors 113, 159, 160–1
effect on services 171, 220
obstructions 37, 50, 71, 160, 213
release of hazards 68, 192
suitability of fills 114–15, 137
pitch 53
planning options 29–30, 71, 103, 131
planning permission 24, 69, 129
plants, hazards to 49, 53, 57, 212
plasterboard 44, 53
plastics 38, 39, 53, 66
plate bearing tests 86–7
police 32, 69, 187
Portland cement 45, 158
preloading 31, 112, 113, 131–2
pressuremeter tests 86–7
private development 205–6
project *see* development
protective clothing 69, 185, 186, 187–8
protective coatings 153–4, 160, 170, 179
Public Health Act 70
public safety 23, 69
pulverised fuel ash (pfa)
as fire-fence 132, 152, 165

pulverised fuel ash (pfa) (*contd*)
  as neutraliser 149
  deposits of 37, 157
  imported fill 132, 157–8, 165
pumping
  gases 127–8, 150–1
  in tests 83, 100
  leachates 129, 155–6
  types of pump 191
  water table level 63, 64, 153–4, 155
purging (borehole) 100, 223–4
putrescible waste
  biodegradation 52, 61–2, 66
  combustion hazard 23
  refuse proportion 38, 39
  site investigation 90, 221
pyritic shales
  aggressive attack 45–7, 95, 101, 123–4
  expansion 45–7, 95, 101, 123–4
  remedial action 123–4, 138, 148

quarries 16, 37, 86, 114, 156

radioactivity
  in pfa 157, 158
  planning constraint 68, 69
  safety 185, 188, 192, 215
  site investigation 79, 101
radiochemist 73
radon gas 158
raft foundations
  design factors 159–60, 160–3
  on fills 31, 131, 171
  on treated crust 123–4, 135, 172, 220
  types of building 115, 135, 159, 172, 220
records 77, 82, 179, 182
redox potential 83, 100, 174
relative stiffness 162–3
remedial treatment 29–30, 103–4
  aggressive ground 147, 153–4, 169–70
  aquifer protection 68, 139, 144
  combustion 151–3, 165
  concrete and sulphates 153–4, 160, 169
  contaminated sites
    bacteria 149
    chemical injection 148
    containment 68, 192
    cover layers 138–45
    encapsulation 154–5
    excavation 32, 145–7, 153–4, 213
    in-situ 147–8
    neutralising 147, 148, 149
    soil processing 148–9
  corrosion 122, 169–70
  cost factors 129, 145, 147, 151, 152
  expansive fills 123–4, 138, 164, 169, 172
  fill sites 31–2, 34, 130–8
  gas
    barriers 142, 143
    drainage 142, 143, 150–1, 166, 168
    extraction 126, 127–8, 150–1, 166
    ventilation 125, 127–8, 150–1, 166–8, 175
  gasworks sites 213
  ground treatment 31–3, 113, 133–4, 137
  landfill gas 149–51
  leachates 129, 147, 154–6
  scrapyards 216
  services 147, 170–2
  underground fires 121, 143, 152–3
rescue equipment 189
resistivity tests 85–6, 219
respiratory equipment 184, 186, 188, 189, 191
restricted areas 182–3, 188–9, 190–1

roads on fills 164–5
rodents 139, 143
root action 57, 65, 139, 143

safety
  asbestos 185–6
  biodegradable fills 187
  biological hazards 185
  building construction 183
  chemical analysis 92–3
  combustibility 28, 151, 152, 183, 186
  contracts 179–80
  contaminants 49, 56–7, 192–3
  contingency plans 189–90
  cost factors 17–18, 71
  demolition 186
  dust and fumes 53, 145, 158
  equipment 184, 185, 186, 187–91
  excavation 145, 186, 188, 189, 191
  explosives 187
  gas 174, 184–5, 188
  health hazards 49, 52, 158
  lagoons 158, 187, 192
  liquid chemicals 184
  maintenance work 182, 183
  pfa 158
  planning constraint 17–18, 71
  radioactivity 185, 188, 192, 215
  rescue equipment 189
  responsibilities 17–18, 24, 69, 179–80, 182
  restricted areas 182–3, 188–9, 190–1
  site facilities 187
  site investigation 27–8, 79, 81, 82, 182–3, 189
  site reclamation 183
  site reconnaissance 182
  solid chemicals 184
  tars 186, 187
  training of staff 190
  voids in fill 183, 187
safety officer 180, 181, 190
safety specialist 22
salinity 148
sampling
  combustibles 90, 101
  contaminants 87–90, 173–4, 211
  depth 89–90
  equipment 97, 98, 100, 101
  expansive fills 95, 101
  gases 87, 89, 100, 127, 174–5, 222–4
  grid pattern 81, 88–9, 210, 211, 215
  groundwater 94, 98, 99–100, 174, 212
  in boreholes 84–5, 94, 98, 99–100, 223–4
  in trial pits 81, 89–90, 97, 99, 211
  liquids 93–4, 98–100
  micro-organisms 101
  protection 81, 83, 88, 98, 100, 101
  radioactivity 101
  slags 95, 101
  solids 93–4, 97–8, 215
  surface 87, 97, 99, 211, 215
  vegetation 86, 101, 173, 215
sampling tubes 83, 97, 100, 174, 224
Scotland 69, 207
scrapyards 74, 91, 143, 213–16
  contaminants in 213–15
sealing plugs 83–5, 98
seasonal effects 35, 63, 84, 94
seismic refraction 85–6
selenium 117, 214
self-heating 58–60, 120, 121
self-ignition 58, 120, 152
self-weight settlement 40–1, 106–7, 108–10, 130, 132
semi-rafts 162

services
  abandoned project 32
  in cover layer 144
  protection of 147, 170–2
  settlement of 160, 168, 220
setting out 82
settlement
  after ground treatment 31, 34, 136
  biodegradable fill 43, 67, 114
  cohesive fills 40, 44, 71, 132, 165
  design factors 159–65
  fill sites 16, 23
  industrial sites 17, 218, 219
  monitoring 75, 77, 81, 165, 175–6
  oil squeeze 65
  planning effects 23, 40, 67, 102, 104
  self-weight 40–1, 106–7, 108–10, 130, 132
  structural loads 41–2, 110–12
  see also collapse settlement; creep settlement;
    differential settlement
sewage sludge 37, 78, 124, 148, 157
sewers
  leachate discharge 70, 129, 156
  protection of 170, 171
  road drainage discharge 164
site appraisal
  acceptance criteria 29, 102
  age of fills 106–7
  chemical attack 121–2
  classification of fills 105, 113–14, 115
    gas hazards 125–6
  combustion 119–21
  compaction of fills 107–8
  construction on fills 104, 113–15
  depth of fills 107
  design factors 30, 114–15
  end use options 29–30, 102–3
  extent of fill 108
  gas emission 78, 102, 124–8
  gasworks sites 212–13
  grading of fill 107
  health hazards 116–19
  landscaping 115–16
  leachates 128–9
  opencast mines 218–19
  remedial works 29–30, 103–4, 115
  scrapyards 216
  settlement of fills 108–13
  slags 123–4
  wastes 123–4
site facilities 82, 180, 187, 193
site history 19, 23, 37, 50, 75
site investigation
  aerial survey 86
  chemical analyses 90–3
  combustion hazard 90, 92, 93, 94–5, 101
  contracts 177
  cost factors 79, 80, 81, 205
  errors in 161
  gases 89, 90, 96, 221–4
  gasworks sites 89, 211
  geophysical methods 85–6
  ground temperatures 94–5
  groundwater 93–4, 98–100, 212
  in-situ geotechnical tests 86–7
  landfill gas 90, 96, 221–4
  leachates 84, 85, 87, 93–4
  migration of contaminants 88, 90, 100
  obstructions 50, 79
  off-site preliminaries 81
  on-site preliminaries 82
  opencast mines 219–20
  preliminary 26–7, 73–80

putrescible waste 90, 221
radioactivity 79, 101
re-evaluation of hazards 29, 71
reports 80, 96
safety 27-8, 79, 81, 82, 182–3, 189
sampling 27–8, 81, 82, 83–5, 87–90
scrapyards 215
slags, spoils, wastes 95
sources of information 50, 76
time required 71
vegetation 86, 101, 173, 215
site laboratory 100
site reconnaissance 27, 77–9, 182, 214
site supervision 181
skin contact 56–7, 158, 184, 186
skip-loading test 87, 111
slag fills 44–5, 49, 50, 95, 123–4
slipcoats 160
smouldering 58–9, 94, 151–2
soil ingestion 54
soil processing 148–9
soil suction
  migration of contaminants 57, 65, 158, 173, 210
  remedial action 139, 143, 144, 154
solvents
  combustion hazards 119
  gas hazard 53
  in drinking water 55, 172
  in groundwater 56, 64
specification 36, 91, 135, 169, 179–80
spent oxide waste
  combustion hazard 58, 60, 119
  cyanides 57, 210
  gasworks residue 209, 210, 211, 213
spiking survey 78, 89, 100, 222–3
standard penetration test 86–7, 219
steelmaking slags
  expansion 45, 95, 101, 123
  obstructions 50
  skin hazard 56
Stockley Park 153, 156, 206
stone columns 31, 133–5, 136, 220
  drainage path 62, 113, 167
  foundations on 135, 163
storage of contaminants 68, 179, 184, 189, 191–2
strip footings
  buildings types 111, 160
  design factors 163
  fill types 115, 159, 172
  treated fills 132, 135
sulphates
  attack on concrete 44, 122, 123
  oxidation process 45–7, 66, 95, 124
  protective measures 153–4, 160, 169, 172
  skin hazard 52, 57
  solubility 56, 121
  trigger values 118
sulphides 53, 57, 118, 211, 215
sulphur dioxide 53, 54
supermarkets 125, 166
superstructures
  design factors 34, 104, 115, 159, 164
  distortion 105, 106, 159, 160
  gas hazard 128
  tilt 159, 162
surcharging 148, 153, 165
Surrey Docks 127, 138, 166–7

target sensitivity 50, 52, 53, 144, T2
tars
  combustion 119
  excavation 145–6, 192
  health hazard 49, 53, 56–7, 186

tars (*contd*)
   migration 65, 67, 139–40, 141
   protection from 186, 187
telephones 69, 170
temperature readings
   grid survey 94, 153, 175
   monitoring 81, 95, 153, 175, 186
   probes 83, 95, 175
tender documents 36, 179, 180–1
Thamesmead site 74, 104, 140, 147
thermal-imaging 78, 86, 95, 175
thermal runaway 58–9
thiocyanates 53, 55, 56, 118
thiosulphates 53
third parties
   leachate migration 68, 128
   protection of 32, 182
   safety responsibilities 17–18, 24, 69, 71, 179
tilt 159, 162
tips, waste
   assessing quantities 89, 147
   special waste 71, 82, 147, 192, 213
   transport to 70, 147
topsoil 65, 157, 158, 173
town gas 60, 209–10
traffic 65, 147, 148, 210
training site staff 190, 191
trenches 42, 70, 83, 150–1, 171
trial pits
   depth of 89–90, 97
   sampling 81, 89–90, 97, 99, 211
trigger values
   fire and explosions 119, 121, 125
   health 102, 116–19, 212, 216

underfloor voids 128, 166–8
underground fires
   excavation 28, 153
   from contaminants 119
   gases produced 53
   putrescible waste 23, 126
   safety 28, 151, 186
   site investigation 94–5
   treatment 121, 143, 152–3
   void settlement 43, 58
undersealing 148, 155
unforeseen hazards 36, 177, 178, 179
Urban Development Grants 24, 206

vanadium 158
vegetables 55, 57, 116, 212, 215, 217
vegetation

as treatment 143, 173
metal hazards 55, 78, 145, 173
site investigation 86, 101, 173, 215
slag hazard 46
ventilation systems 125, 127–8, 150–1, 166–8, 175
   trench 127, 150, 151
   voids 158, 166–8, 175
vibratory compaction 31, 62, 115, 133–6, 220
voids, ventilation 158, 166–8, 175
voids in fills
   biodegradation 43, 171, 187
   combustion 38, 58
   gas migration 125
   safety 183, 187
   settlement 38, 43, 58

Wales, Survey of Contaminated Land 71
warehouses 125, 138, 166, 168, 169
waste disposal
   leachates 129, 156
   on site 147, 189, 213, 216
   pumped extraction 127, 156
   responsibilities 71, 93, 147
   special waste 82, 147, 192, 213, 216
   transport 70, 147
water authority 32, 69, 179
   leachates 38, 129, 154
water drainage layer 142, 143, 144
water gas 152, 186
water supplies 70, 154, 170
water supplies, hazards to 16, 32, 102
   infiltration barriers 139, 154–6
   leachates 23, 38, 63, 70, 93, 128–9
   monitoring 174, 212
   phenols and solvents 55–6, 172
   piling 68
water table levels
   contaminant hazard 17, 52, 129
   cover layers 140
   deep drains 63, 129, 153, 155
   monitoring 83, 174
   pumping 63, 64, 153–4, 155
   settlement 42, 84, 112–13, 161, 218
wells
   contamination 55
   gas drainage 150, 151
   leachates 64, 129, 155

zinc 119, 217
   phytotoxicity 57, 116, 117, 212, 215
   scrapyards 214, 215